Chapter 3
Understand Multiplication

Made in the United States
Text printed on 100%
recycled paper

Houghton
Mifflin
Harcourt

Printed in the U.S.A.

ISBN 978-0-544-34209-5

9 10 0928 18 17 16

4500602565 B C D E F G

Dear Students and Families,

Welcome to **Go Math!**, Grade 3! In this exciting mathematics program, there are hands-on activities to do and real-world problems to solve. Best of all, you will write your ideas and answers right in your book. In **Go Math!**, writing and drawing on the pages helps you think deeply about what you are learning, and you will really understand math!

By the way, all of the pages in your **Go Math!** book are made using recycled paper. We wanted you to know that you can Go Green with **Go Math!**

Sincerely,

The Authors

Made in the United States
Text printed on 100% recycled paper

Authors

Juli K. Dixon, Ph.D.
Professor, Mathematics Education
University of Central Florida
Orlando, Florida

Edward B. Burger, Ph.D.
President, Southwestern University
Georgetown, Texas

Steven J. Leinwand
Principal Research Analyst
American Institutes for Research (AIR)
Washington, D.C.

Contributor

Rena Petrello
Professor, Mathematics
Moorpark College
Moorpark, California

Matthew R. Larson, Ph.D.
K-12 Curriculum Specialist for Mathematics
Lincoln Public Schools
Lincoln, Nebraska

Martha E. Sandoval-Martinez
Math Instructor
El Camino College
Torrance, California

English Language Learners Consultant

Elizabeth Jiménez
CEO, GEMAS Consulting
Professional Expert on English Learner Education
Bilingual Education and Dual Language
Pomona, California

Whole Number Operations

Critical Area Developing understanding of multiplication and division and strategies for multiplication and division within 100

Understand Multiplication 137

COMMON CORE STATE STANDARDS

3.OA Operations and Algebraic Thinking

Cluster A Represent and solve problems involving multiplication and division.
3.OA.A.1, 3.OA.A.3

Cluster B Understand properties of multiplication and the relationship between multiplication and division.
3.OA.B.5

Cluster D Solve problems involving the four operations, and identify and explain patterns in arithmetic.
3.OA.D.8

Critical Area

GO DIGITAL

Go online! Your math lessons are interactive. Use *i*Tools, Animated Math Models, the Multimedia *e*Glossary, and more.

Chapter 3 Overview

In this chapter, you will explore and discover answers to the following **Essential Questions**:

- How can you use multiplication to find how many in all?
- What models can help you multiply?
- How can you use skip counting to help you multiply?
- How can multiplication properties help you find products?
- What types of problems can be solved by using multiplication?

Personal Math Trainer
Online Assessment and Intervention

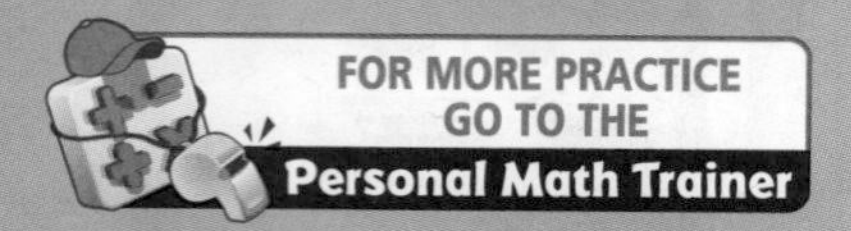

Practice and Homework

Lesson Check and Spiral Review in every lesson

Chapter 3

Understand Multiplication

Check your understanding of important skills.

Name ________________________________

▶ Count On to Add **Use the number line. Write the sum.** (1.OA.C.5)

1. $6 + 2 =$ _____

2. $3 + 7 =$ _____

▶ Skip Count by Twos and Fives **Skip count. Write the missing numbers.** (2.NBT.A.2)

3. 2, 4, 6, _____, _____, _____

4. 5, 10, 15, _____, _____, _____

▶ Model with Arrays **Use the array. Complete.** (2.OA.C.4)

5.

____ + ____ + ____ = ____

6.

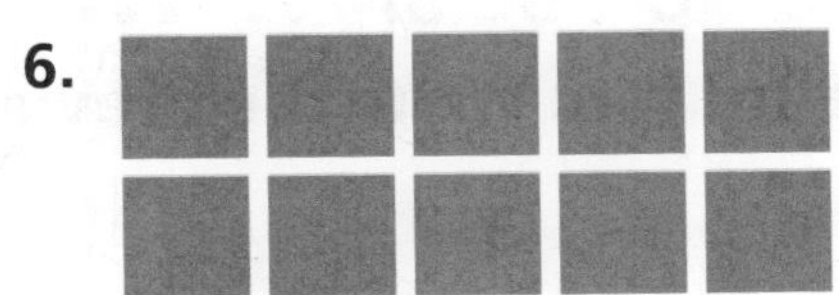

____ + ____ = ____

Math in the Real World

Ryan's class went on a field trip to a farm. They saw 5 cows and 6 chickens. Help to find how many legs were on all the animals they saw.

Vocabulary Builder

▶ Visualize It

Complete the tree map by using the review words.

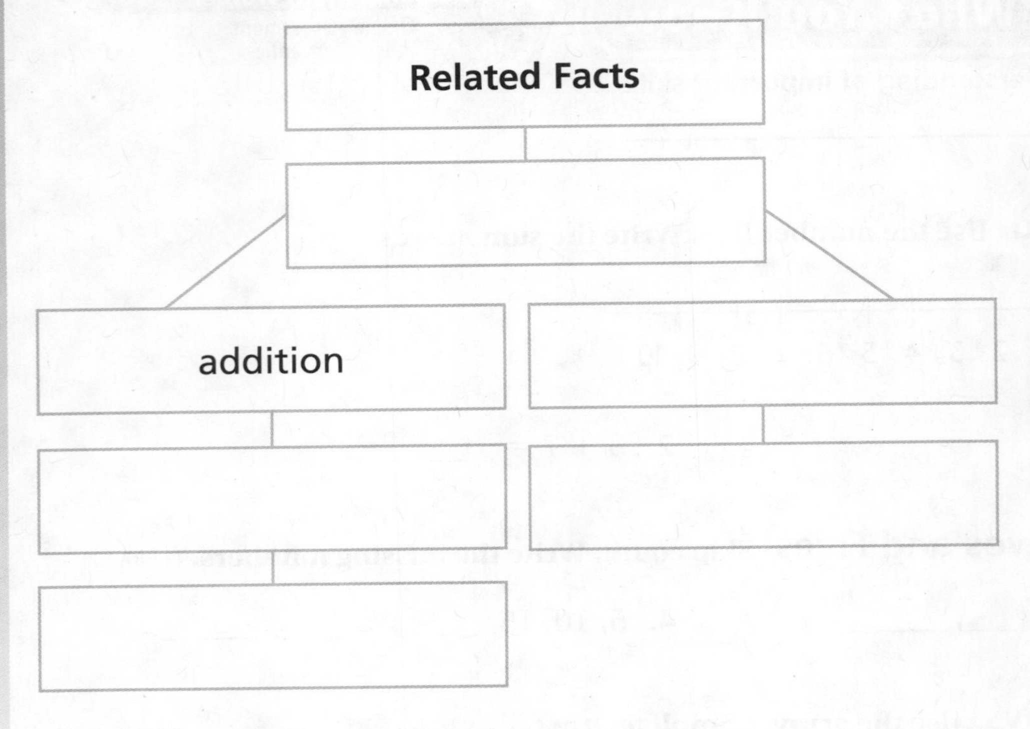

Review Words
addend
addition
difference
number sentences
related facts
subtraction
sum

Preview Words
array
equal groups
factor
multiply
product

▶ Understand Vocabulary

Read the definition. Write the preview word that matches it.

1. A set of objects arranged in rows and columns ____________

2. The answer in a multiplication problem ____________

3. When you combine equal groups to find how many in all ____________

4. A number that is multiplied by another number to find a product ____________

• Interactive Student Edition
• Multimedia eGlossary

Chapter 3 Vocabulary

array

matriz

4

Commutative Property of Multiplication

Propiedad conmutativa de la multiplicación

9

equal groups

grupos iguales

20

factor

factor

25

Identity Property of Multiplication

Propiedad de identidad de la multiplicación

35

multiply

multiplicar

51

product

producto

65

Zero Property of Multiplication

Propiedad del cero de la multiplicación

85

The property that states that you can multiply two factors in any order and get the same product

Example: $4 \times 3 = 3 \times 4$

A set of objects arranged in rows and columns

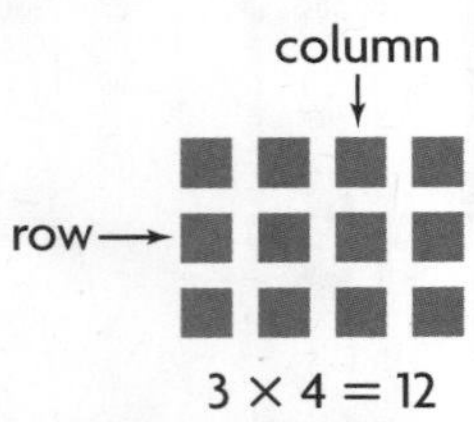

A number that is multiplied by another number to find a product

Example: $4 \times 5 = 20$

factor factor

Groups that have the same number of objects

To combine equal groups to find how many in all; the opposite operation of division

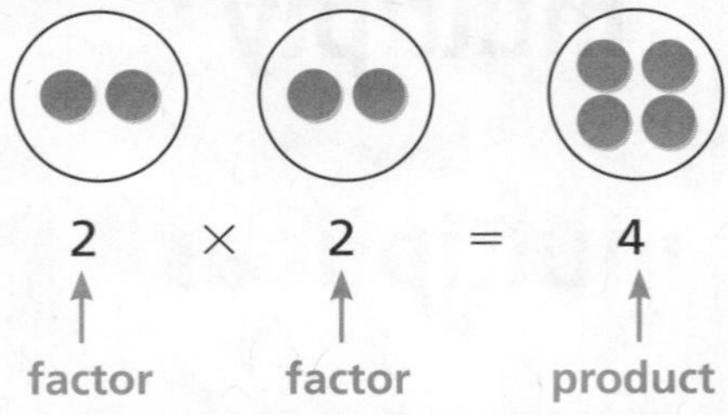

The property that states that the product of any number and 1 is that number

Example: $17 \times 1 = 17$

The property that states that the product of zero and any number is zero

Example: $34 \times 0 = 0$

The answer in a multiplication problem

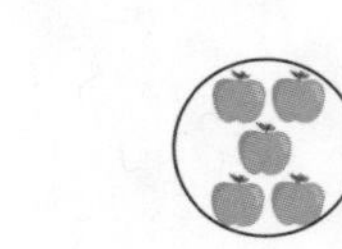   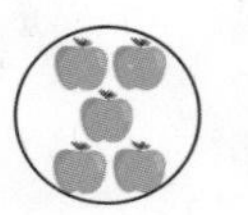

Example: $4 \times 5 = 20$

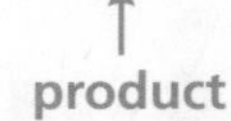

Matchup

Word Box
array
equal groups
factors
multiply
product
Commutative Property of Multiplication
Identity Property of Multiplication
Zero Property of Multiplication

For 2–3 players

Materials

1 set of word cards

How to Play

1. Put the cards face-down in rows. Take turns to play.
2. Choose two cards and turn them face-up.
   - If the cards show a word and its meaning. it's a match. Keep the pair and take another turn.
   - If the cards do not match, turn them back over.
3. The game is over when all cards have been matched. The players count their pairs. The player with the most pairs wins.

The Write Way

Reflect

Choose one idea. Write about it.

- Do $4 + 4 + 4$ and 4×3 represent equal groups? Explain why or why not.
- Explain how to use an array to find a product.
- Summarize how to solve 5×0, including any "false starts" or "dead ends" you might take.

Name ______________________________

Count Equal Groups

Essential Question How can you use equal groups to find how many in all?

Common Core **Operations and Algebraic Thinking— 3.OA.A.1** *Also 3.OA.A.3*
MATHEMATICAL PRACTICES
MP2, MP3, MP4

Unlock the Problem

Hands On

Equal groups have the same number of objects in each group.

Tim has 6 toy cars. Each car has 4 wheels. How many wheels are there in all?

- How many wheels are on each car?

  ____

- How many equal groups of wheels are there?

  ____

- How can you find how many wheels in all?

  ____

Activity Use counters to model the equal groups.

Materials ■ counters

STEP 1 Draw 4 counters in each group.

STEP 2 Skip count to find how many wheels in all. Skip count by 4s until you say 6 numbers.

number of equal groups → 1 2 3 4 5 6

4, ____, 12, ____, ____, ____

There are ____ groups with ____ wheels in each group.

So, there are ____ wheels in all.

MATHEMATICAL PRACTICES 2

Reason Quantitatively What if Tim had 8 cars? How could you find the total number of wheels?

Example Count equal groups to find the total.

Sam, Kyla, and Tia each have 5 pennies.
How many pennies do they have in all?

How many pennies does each person have? ____________

How many equal groups of pennies are there? ____________

Draw 5 counters in each group.

Think: There are _____ groups of 5 pennies.

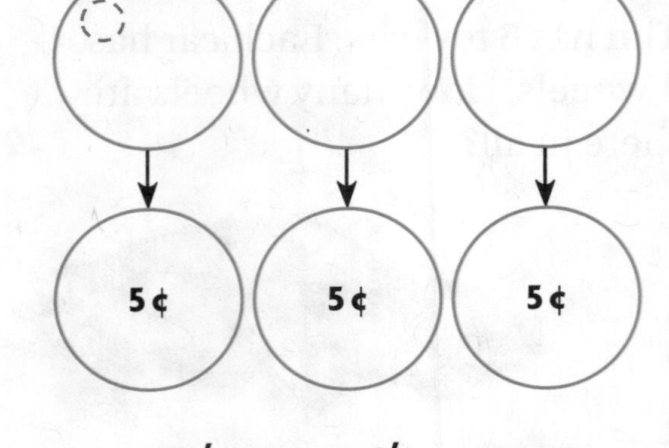

Think: There are _____ fives.

Skip count to find how many pennies. _____, _____, _____

So, they have _____ pennies.

- THINK SMARTER Explain why you can skip count by 5s to find how many.

____

Share and Show

1. Complete. Use the picture. Skip count to find how many wheels in all.

_____ groups of 2

_____ twos

Skip count by 2s. 2, 4, _____, _____

So, there are _____ wheels.

Math Talk MATHEMATICAL PRACTICES 3

Apply How would your answer change if 2 more groups of wheels were added?

Name ______________________

Draw equal groups. Skip count to find how many.

2. 2 groups of 6 _____

3. 3 groups of 2 _____

Count equal groups to find how many.

4. 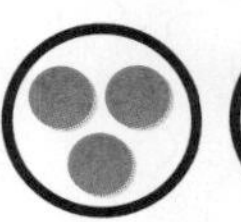 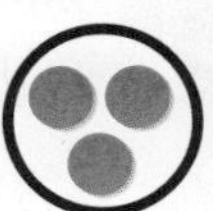  

_____ groups of _____

_____ in all

5. 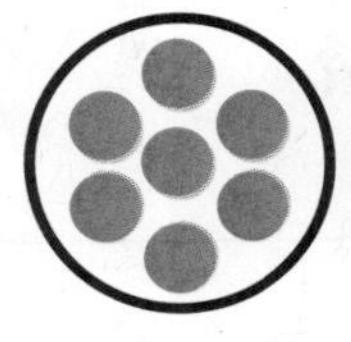 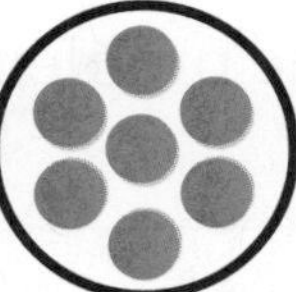 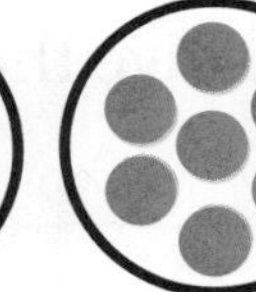

_____ groups of _____

_____ in all

On Your Own

Draw equal groups. Skip count to find how many.

6. 3 groups of 3 _____

7. 2 groups of 9 _____

8. GO DEEPER A toy car costs $3. A toy truck costs $4. Which costs more—4 cars or 3 trucks? Explain.

____

____

9. MATHEMATICAL PRACTICE 3 **Make Arguments** Elliott has a collection of 20 toy cars. Will he be able to put an equal number of toy cars on 3 shelves? Explain your answer.

____

____

Unlock the Problem

10. THINKSMARTER Tina, Charlie, and Amber have toy cars. Each car has 4 wheels. How many wheels do their cars have altogether?

a. What do you need to find?

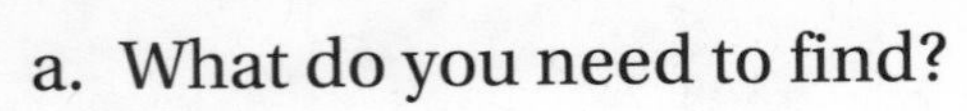

b. What information will you use from the graph to solve the problem?

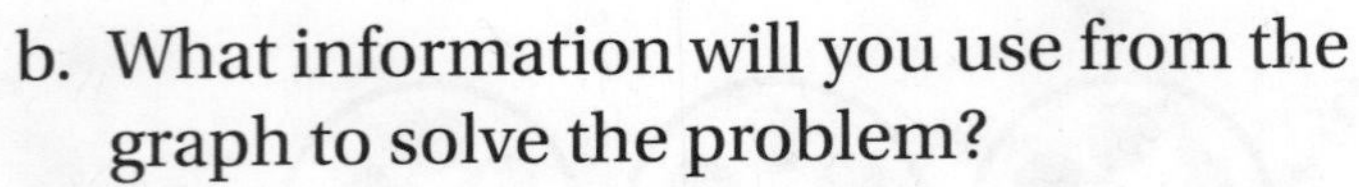

c. Show the steps you used to solve the problem.

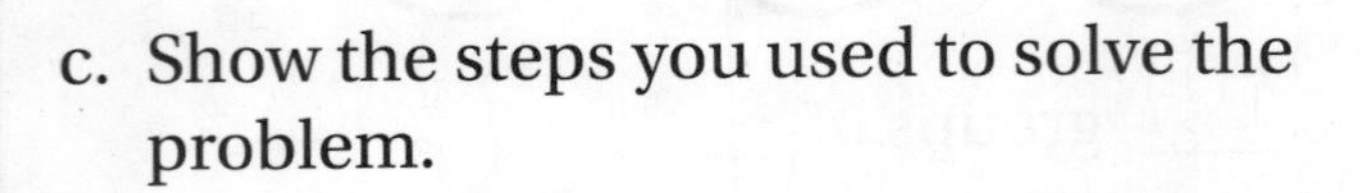

d. So, the cars have _____ wheels.

11. THINKSMARTER A bookcase has 4 shelves. Each shelf holds 5 books. How many books are in the bookcase?

Draw counters to model the problem. Then explain how you solved the problem.

____

____

Name ____________________

Count Equal Groups

COMMON CORE STANDARD—3.OA.A.1
Represent and solve problems involving multiplication and division.

Draw equal groups. Skip count to find how many.

1. 2 groups of 2 __4__

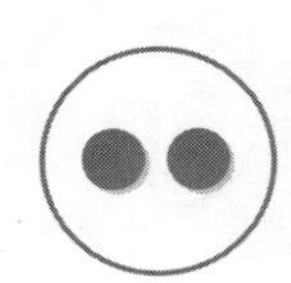 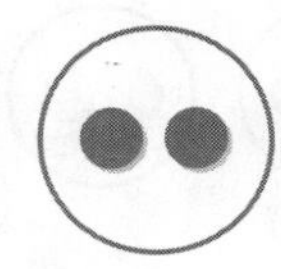

2. 3 groups of 6 ______

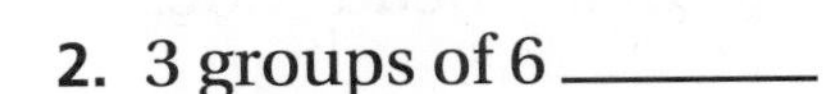

Count equal groups to find how many.

3. 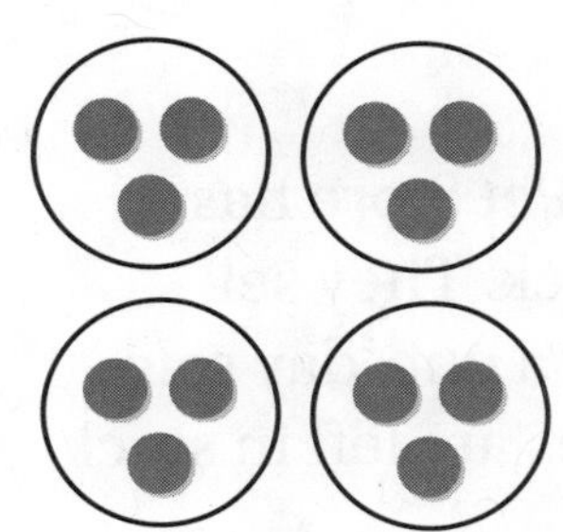

______ groups of ______

______ in all

4. 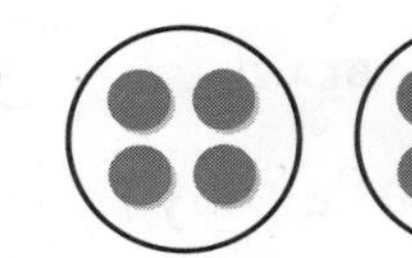
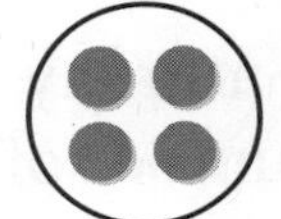 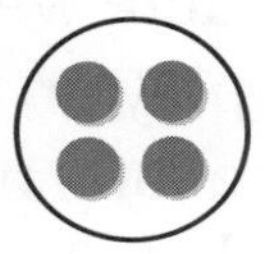

______ groups of ______

______ in all

Problem Solving (Real World)

5. Marcia puts 2 slices of cheese on each sandwich. She makes 4 cheese sandwiches. How many slices of cheese does Marcia use in all?

____

6. Tomas works in a cafeteria kitchen. He puts 3 cherry tomatoes on each of 5 salads. How many tomatoes does he use?

____

7. **WRITE** *Math* Write a problem that can be solved by using equal groups.

____

____

Lesson Check (3.OA.A.1)

1. Jen makes 3 bracelets. Each bracelet has 3 beads. How many beads does Jen use?

 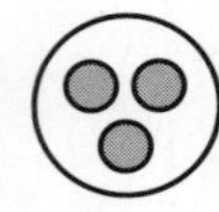 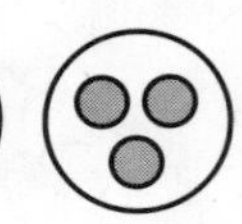

____

2. Ian has 5 cards to mail. Each card needs 2 stamps. How many stamps does Ian need?

____

Spiral Review (3.NBT.A.1, 3.NBT.A.2)

3. There were 384 people at a play on Friday night. There were 512 people at the play on Saturday night. Estimate the total number of people who attended the play on both nights.

____

____

4. Walking the Dog Pet Store has 438 leashes in stock. They sell 79 leashes during a one-day sale. How many leashes are left in stock after the sale?

____

____

5. The Lakeside Tour bus traveled 490 miles on Saturday and 225 miles on Sunday. About how many more miles did it travel on Saturday?

____

____

6. During one week at Jackson School, 210 students buy milk and 196 students buy juice. How many drinks are sold that week?

____

____

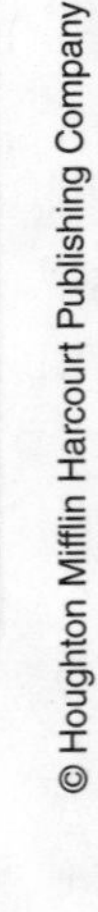

FOR MORE PRACTICE GO TO THE **Personal Math Trainer**

Name ________________________________

Relate Addition and Multiplication

Essential Question How is multiplication like addition? How is it different?

Operations and Algebraic Thinking—3.OA.A.1 *Also 3.OA.A.3, 3.OA.C.7, 3.NBT.A.2*

MATHEMATICAL PRACTICES
MP3, MP4, MP6

Unlock the Problem

Tomeka needs 3 apples to make one loaf of apple bread. Each loaf has the same number of apples. How many apples does Tomeka need to make 4 loaves?

- How many loaves is Tomeka making?

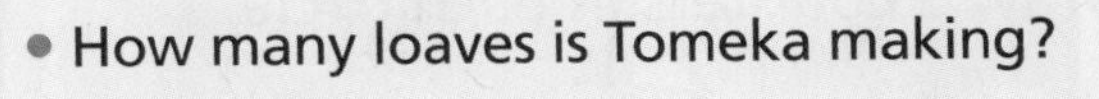

- How many apples are in each loaf?

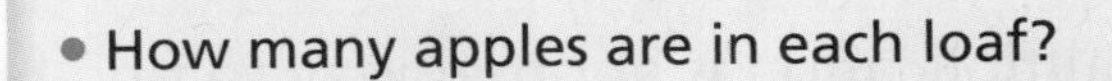

- How can you solve the problem?

____

One Way Add equal groups.

Use the 4 circles to show the 4 loaves.

Draw 3 counters in each circle to show the apples Tomeka needs for each loaf.

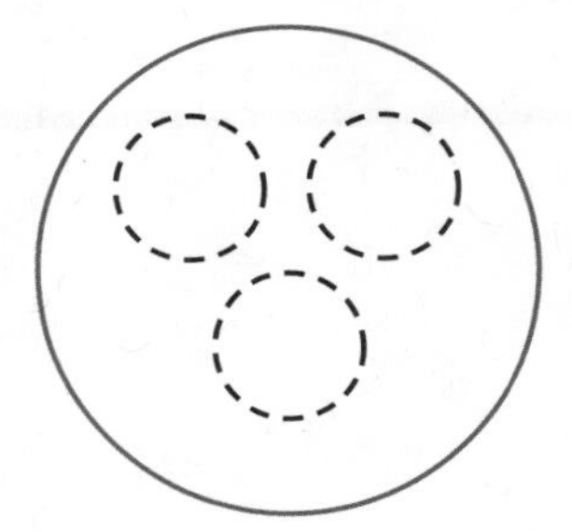

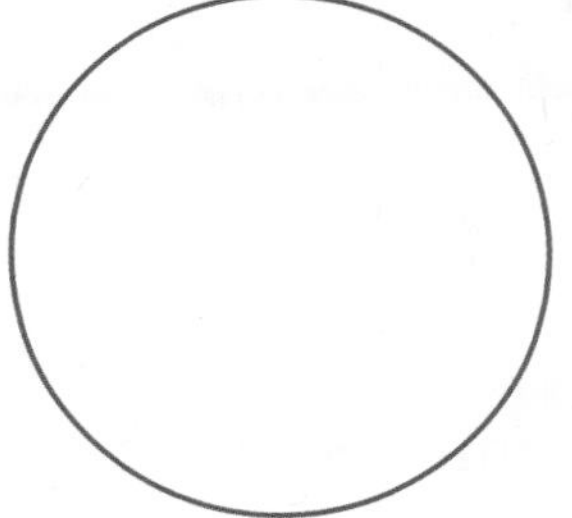

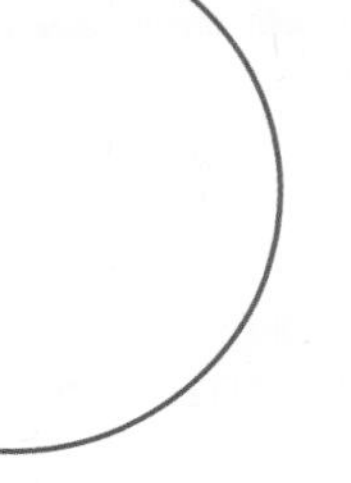

Find the number of counters.
Complete the addition sentence.

3 + _____ + _____ + _____ = _____

So, Tomeka needs _____ apples to

make _____ loaves of apple bread.

Math Talk MATHEMATICAL PRACTICES 4

Use Diagrams How can drawing a picture help you to solve a multiplication problem?

Another Way Multiply.

When you combine equal groups, you can **multiply** to find how many in all.

Think: 4 groups of 3

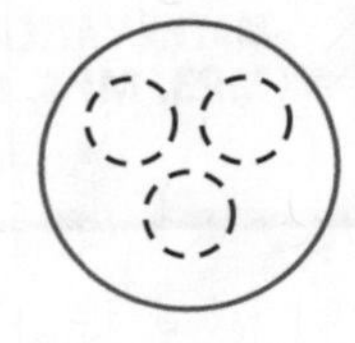
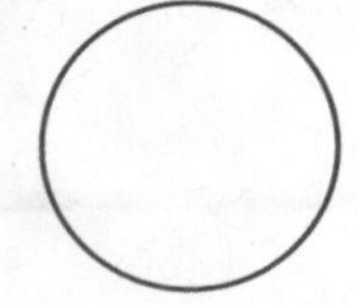
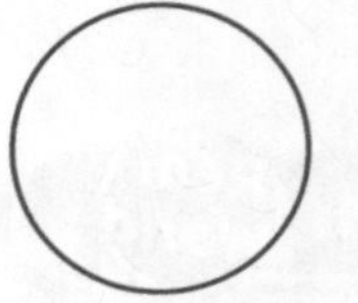
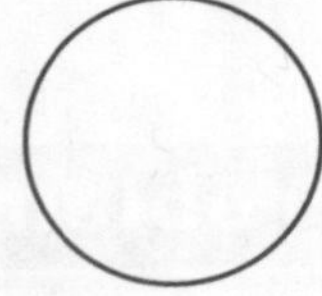

Draw 3 counters in each circle.

Since there are the same number of counters in each circle, you can multiply to find how many in all.

Multiplication is another way to find how many there are altogether in equal groups.

Write: $4 \times 3 = 12$ (4 ← factor, 3 ← factor, 12 ← product)

or

$$\begin{array}{r} 4 \\ \times\ 3 \\ \hline 12 \end{array}$$

4 ← factor
× 3 ← factor
12 ← product

Read: Four times three equals twelve.

The **factors** are the numbers multiplied.

The **product** is the answer to a multiplication problem.

Share and Show

1. Write related addition and multiplication sentences for the model.

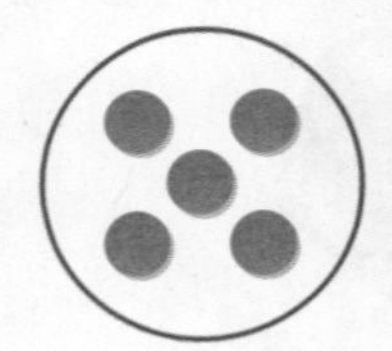
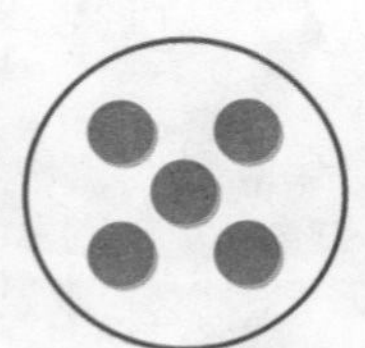
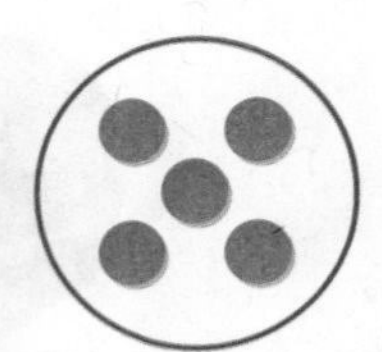
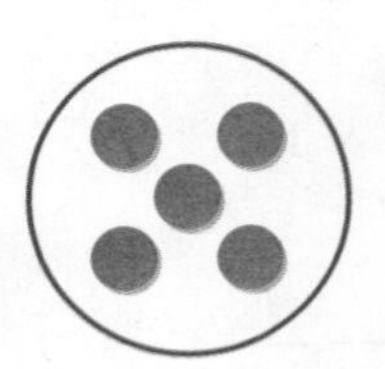

_____ + _____ + _____ + _____ = _____

_____ × _____ = _____

MATHEMATICAL PRACTICES 4

Use Models How would you change this model so you could write a multiplication sentence to match it?

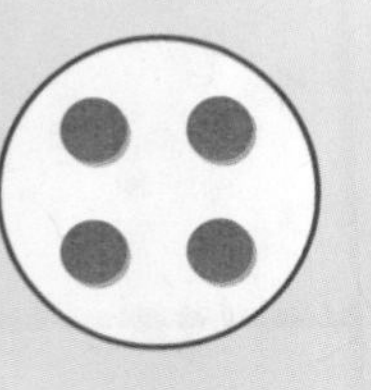
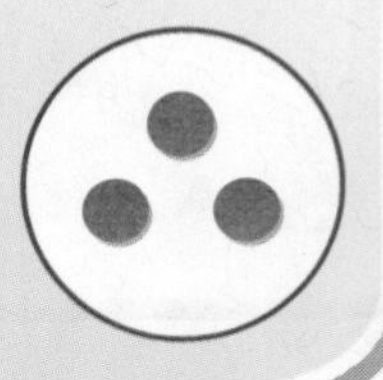

Name ____________________

Draw a quick picture to show the equal groups. Then write related addition and multiplication sentences.

2. 3 groups of 6

___ + ___ + ___ = ___

___ × ___ = ___

3. 2 groups of 3

___ + ___ = ___

___ × ___ = ___

On Your Own

Draw a quick picture to show the equal groups. Then write related addition and multiplication sentences.

4. 4 groups of 2

___ + ___ + ___ + ___ = ___

___ × ___ = ___

5. 5 groups of 4

___ + ___ + ___ + ___ + ___ = ___

___ × ___ = ___

Complete. Write a multiplication sentence.

6. Zach buys 4 packs of pens. Each pack has 4 pens. Write a multiplication sentence to show how many pens Zach buys.

___ × ___ = ___

7. Ada has 3 vases. She puts 5 flowers in each vase. Write a multiplication sentence to show how many flowers Ada puts in the vases.

___ × ___ = ___

8. GO DEEPER Mrs. Tomar buys 2 packs of vanilla yogurt and 3 packs of strawberry yogurt. Each pack has 4 yogurts. How many yogurts does Mrs. Tomar buy?

____

9. GO DEEPER Murray buys 3 packs of red peppers and 4 packs of green peppers. Each pack has 4 peppers. How many peppers does Murray buy?

____

Problem Solving • Applications

Use the table for 10–11.

Fruit	Weight in Ounces
Apple	6
Orange	5
Peach	3
Banana	4

Average Weight of Fruits

10. Morris bought 4 peaches. How much do the peaches weigh? Write a multiplication sentence to find the weight of the peaches.

_____ $\times$ _____ $=$ _____ ounces

11. THINK SMARTER Thomas bought 2 apples. Sydney bought 4 bananas. Which weighed more—the 2 apples or the 4 bananas? How much more? Explain how you know.

____

____

____

12. MATHEMATICAL PRACTICE 3 **Make Arguments** Shane said that he could write related multiplication and addition sentences for $6 + 4 + 3$. Does Shane's statement make sense? Explain.

____

____

13. GO DEEPER Write a word problem that can be solved using 3×4. Solve the problem.

____

____

14. THINK SMARTER Select the number sentences that represent the model at the right. Mark all that apply.

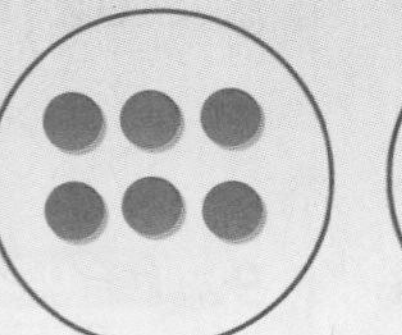 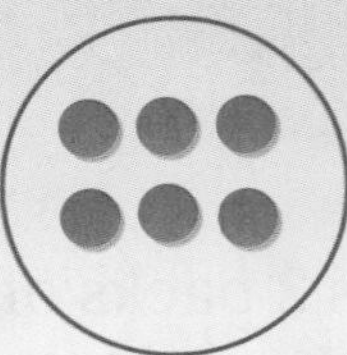 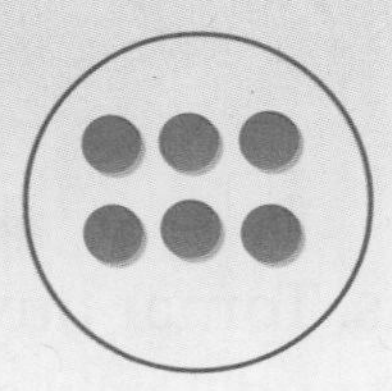

Ⓐ $3 + 6 = 9$

Ⓑ $6 + 6 + 6 = 18$

Ⓒ $3 \times 6 = 18$

Ⓓ $6 + 3 = 9$

Name ____________________

Relate Addition and Multiplication

Common Core

COMMON CORE STANDARD—3.OA.A.1
Represent and solve problems involving multiplication and division.

Draw a quick picture to show the equal groups. Then write related addition and multiplication sentences.

1. 3 groups of 5

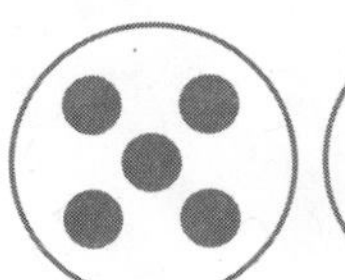 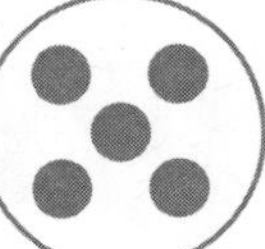 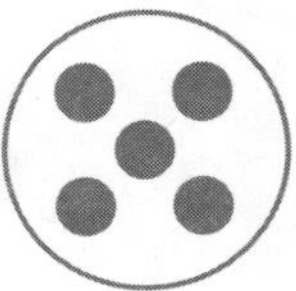

 5 + 5 + 5 = 15

 3 × 5 = 15

2. 3 groups of 4

 ___ + ___ + ___ = ___

 ___ × ___ = ___

3. 5 groups of 2

 ___ + ___ + ___ + ___ + ___ = ___

 ___ × ___ = ___

Complete. Write a multiplication sentence.

4. 7 + 7 + 7 = ___

 ___ × ___ = ___

5. 3 + 3 + 3 = ___

 ___ × ___ = ___

Problem Solving

6. There are 6 jars of pickles in a box. Ed has 3 boxes of pickles. How many jars of pickles does he have? Write a multiplication sentence to find the answer.

 ___ × ___ = ___ jars

7. Each day, Jani rides her bike 5 miles. How many miles does Jani ride in 4 days? Write a multiplication sentence to find the answer.

 ___ × ___ = ___ miles

8. **WRITE** ▸ *Math* Write a word problem that involves combining three equal groups.

____

____

Lesson Check (3.OA.A.1)

1. What is another way to show

   $3 + 3 + 3 + 3 + 3 + 3$?

2. Use the model. How many counters are there?

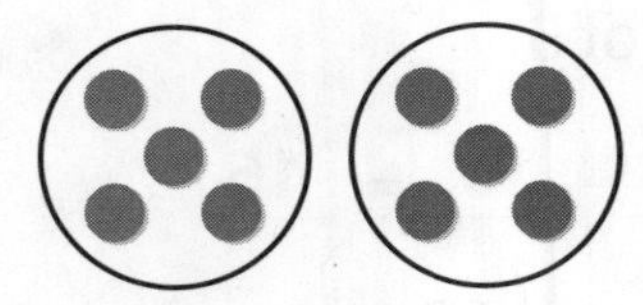

Spiral Review (3.NBT.A.1, 3.NBT.A.2, 3.MD.B.4)

3. A school gave 884 pencils to students on the first day of school. What is 884 rounded to the nearest hundred?

4. Find the difference.

$$\begin{array}{r} 632 \\ -\ 274 \\ \hline \end{array}$$

5. The line plot below shows how many points Trevor scored in 20 games.

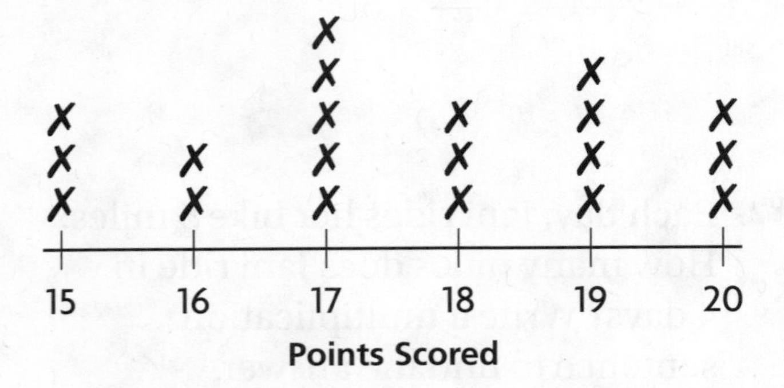

   In how many games did Trevor score 18 points or fewer?

6. Darrien read 97 pages last week. Evan read 84 pages last week. How many pages did the boys read?

FOR MORE PRACTICE GO TO THE Personal Math Trainer

Name ____________________________________

Skip Count on a Number Line

Essential Question How can you use a number line to skip count and find how many in all?

Common Core **Operations and Algebraic Thinking— 3.OA.A.3** *Also 3.OA.A.1*

MATHEMATICAL PRACTICES
MP3, MP4, MP7

Unlock the Problem

Caleb wants to make 3 balls of yarn for his cat to play with. He uses 6 feet of yarn to make each ball. How many feet of yarn does Caleb need in all?

- How many equal groups of yarn will Caleb make?

  ____

- How many feet of yarn will be in each group?

  ____

- What do you need to find?

  ____

Use a number line to count equal groups.

How many feet of yarn does Caleb need for each ball? ________

How many equal lengths of yarn does he need? ________

Begin at 0. Skip count by 6s by drawing jumps on the number line.

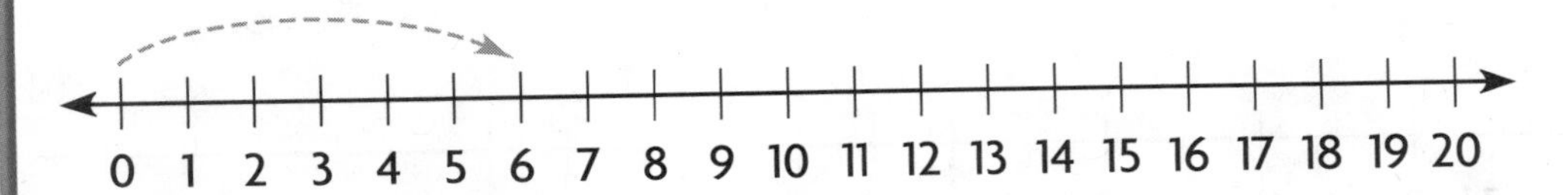

How many jumps did you make? ________

How long is each jump? ________

Multiply. $3 \times 6 =$ ______

So, Caleb needs ______ feet of yarn in all.

MATHEMATICAL PRACTICES 3

Compare Representations How would what you draw on the number line change if instead of 3 balls of yarn made with 6 feet of yarn there were 4 balls of yarn made with 5 feet of yarn?

- MATHEMATICAL PRACTICE 1 **Analyze** Why did you jump by 6s on the number line?

  ____

Share and Show MATH BOARD

1. Skip count by drawing jumps on the number line. Find how many in 5 jumps of 4. Then write the product.

Think: 1 jump of 4 shows 1 group of 4.

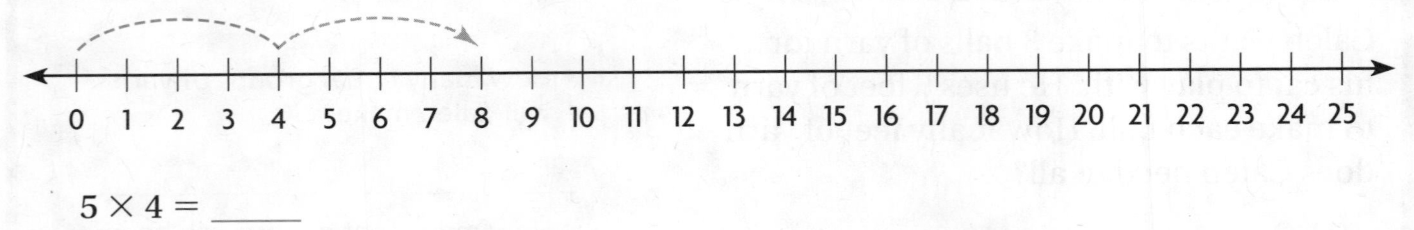

$5 \times 4 =$ _____

Draw jumps on the number line to show equal groups. Find the product.

2. 3 groups of 8

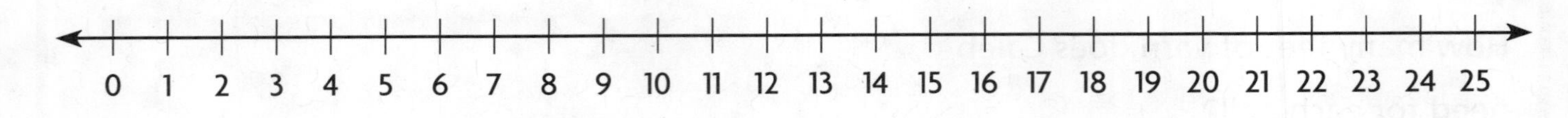

$3 \times 8 =$ _____

3. 8 groups of 3

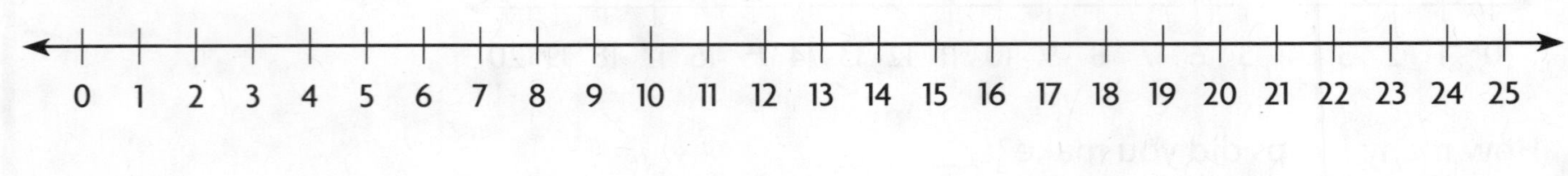

$8 \times 3 =$ _____

Write the multiplication sentence shown by the number line.

4.

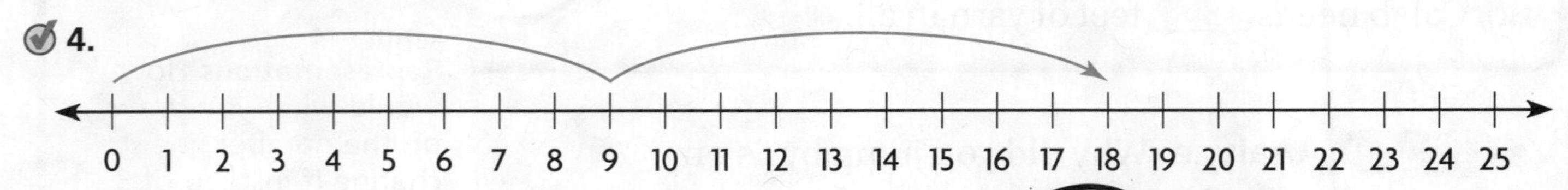

_____ $\times$ _____ $=$ _____

Math Talk — MATHEMATICAL PRACTICES 4

Model Mathematics How do equal jumps on the number line show equal groups?

Name ___________________________

On Your Own

Draw jumps on the number line to show equal groups. Find the product.

5. 6 groups of 4

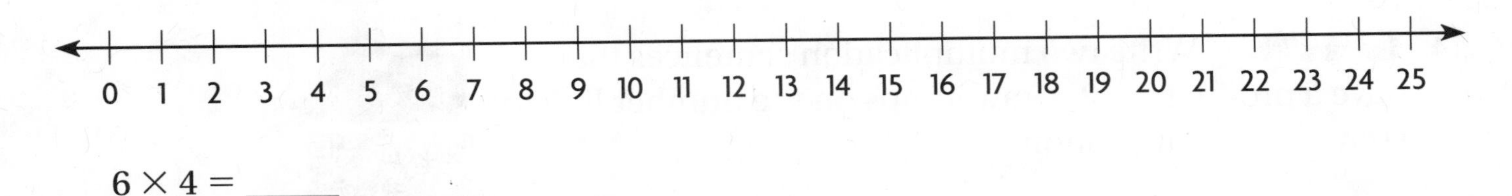

$6 \times 4 =$ _____

6. 7 groups of 3

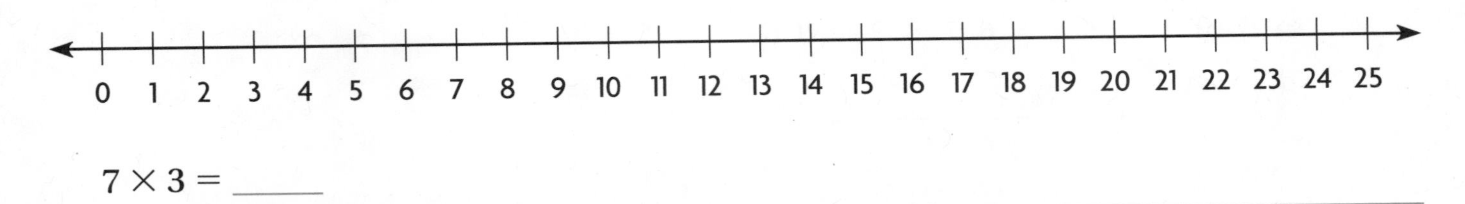

$7 \times 3 =$ _____

7. Sam, Kyra, Tia, and Abigail each have 10 pennies. How many pennies do they have in all?

____

8. Eddie bought snacks for a picnic. He has 3 bags of snacks. Each bag has 4 snacks. How many snacks does Eddie have in all?

____

9. Ashley digs 7 holes. She puts 2 seeds in each hole. She has 3 seeds left over. How many seeds are there in all?

____

10. GO DEEPER Carla puts 8 pictures on each page of a photo album. She fills 3 pages. She has 5 pictures left. How many pictures does she have?

____

11. GO DEEPER A band marches in rows of 5. Each row has 6 people. There are 4 people who carry flags. How many people are in the marching band?

____

12. GO DEEPER In Mr. Gupta's classroom, there are 4 rows of desks. Each row has 6 desks. Mrs. Loew's classroom has 3 rows of 9 desks. How many desks are in Mr. Gupta's and Mrs. Loew's classrooms?

____

Problem Solving • Applications

13. GO DEEPER Erin displays her toy cat collection on 3 shelves. She puts 8 cats on each shelf. If she collects 3 more cats, how many cats will she have?

____

14. THINK SMARTER Write two multiplication sentences that have a product of 12. Draw jumps on the number line to show the multiplication.

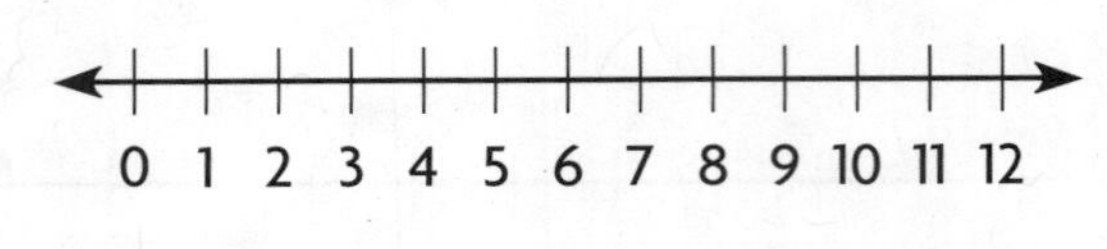

____ × ____ = ____

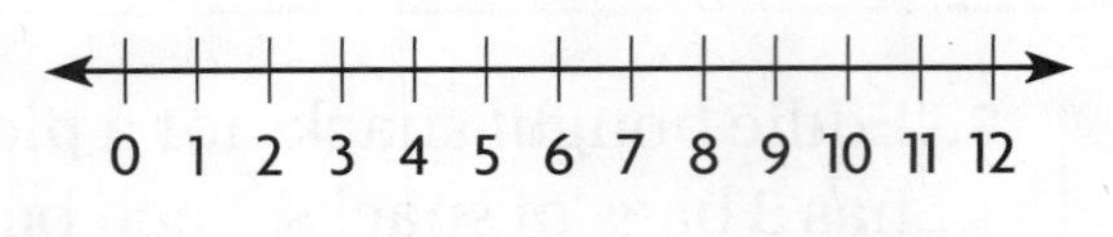

____ × ____ = ____

15. MATHEMATICAL PRACTICE 7 **Identify Relationships** Write a problem that can be solved by finding 8 groups of 5. Write a multiplication sentence to solve the problem. Then solve.

____

____

Personal Math Trainer

16. THINK SMARTER + Rebecca practices piano for 3 hours each week. How many hours does she practice in 4 weeks?

Draw jumps and label the number line to show your thinking.

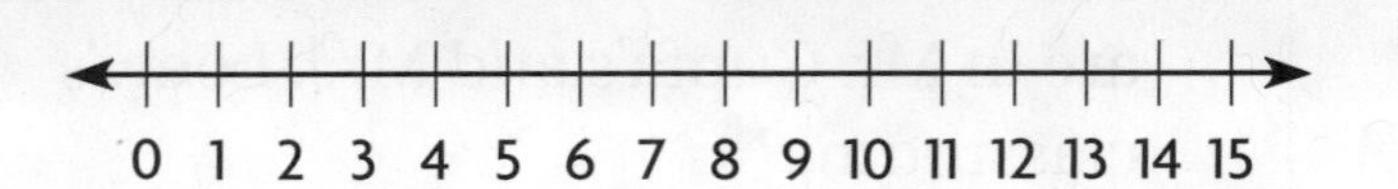

____

Name ______________________________

Skip Count on a Number Line

Common Core

COMMON CORE STANDARD—3.OA.A.3
Represent and solve problems involving multiplication and division.

Draw jumps on the number line to show equal groups. Find the product.

1. 6 groups of 3

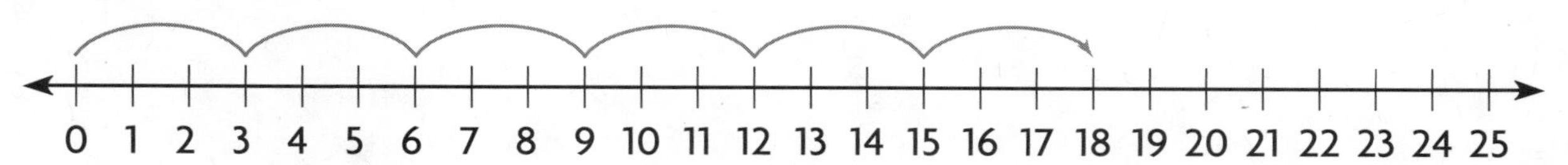

$6 \times 3 =$ 18

Write the multiplication sentence the number line shows.

2. 2 groups of 6

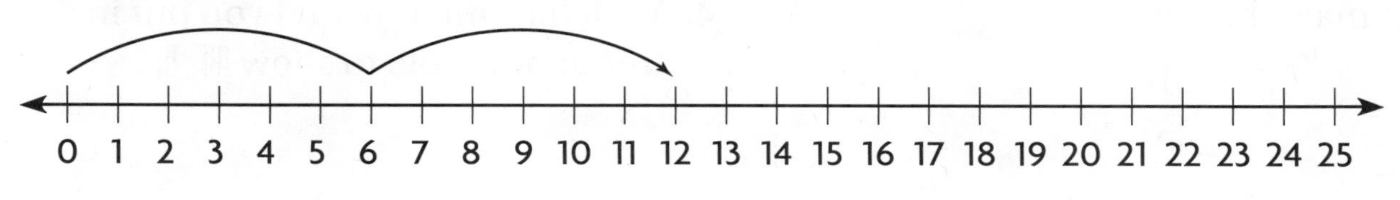

_______ × _______ = _______

Problem Solving Real World

3. Allie is baking muffins for students in her class. There are 6 muffins in each baking tray. She bakes 5 trays of muffins. How many muffins is she baking?

____

4. A snack package has 4 cheese sticks. How many cheese sticks are in 4 packages?

____

5. **WRITE** *Math* Write a problem that can be solved by skip counting on a number line.

____

____

Lesson Check (3.OA.A.3)

1. Louise skip counts by 4 on a number line to find 5×4. How many jumps should she draw on the number line?

2. Theo needs 4 boards that are each 3 feet long to make bookshelves. How many feet of boards does he need altogether?

Spiral Review (3.NBT.A.1, 3.MD.B.3)

3. Estimate the sum.

$$\begin{array}{r} 518 \\ +251 \\ \hline \end{array}$$

4. Which number would you put in a frequency table to show 𝍸 ||| ?

5. A manager at a shoe store received an order for 346 pairs of shoes. What is 346 rounded to the nearest hundred?

6. Toby is making a picture graph. Each picture of a book is equal to 2 books he has read. The row for Month 1 has 3 pictures of books. How many books did Toby read during Month 1?

FOR MORE PRACTICE GO TO THE Personal Math Trainer

Name ______________________________

Vocabulary

Choose the best term from the box.

Vocabulary
equal groups
factors
multiply
product

1. When you combine equal groups, you can ______________ to find how many in all. (p. 146)
2. The answer in a multiplication problem is called the ______________. (p. 146)
3. The numbers you multiply are called the ______________. (p. 146)

Concepts and Skills

Count equal groups to find how many. (3.OA.A.1)

4. 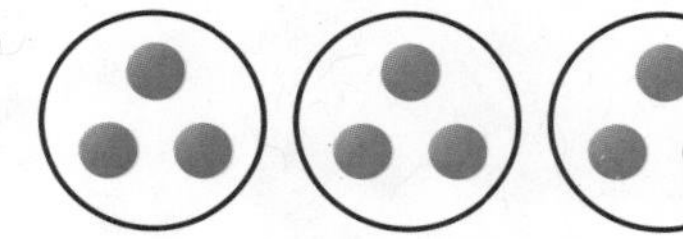

___ groups of ___

___ in all

5. 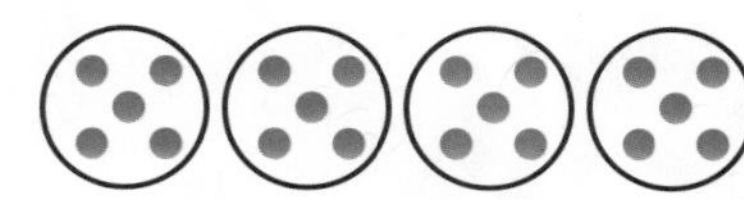

___ groups of ___

___ in all

6. 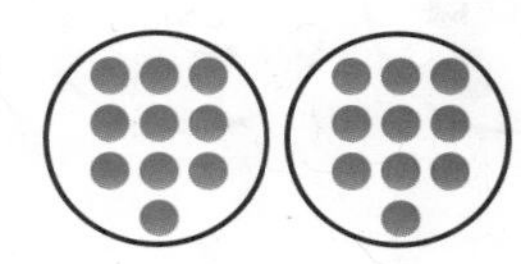

___ groups of ___

___ in all

Write related addition and multiplication sentences. (3.OA.A.1)

7. 3 groups of 9

___ + ___ + ___ = ___

___ × ___ = ___

8. 5 groups of 7

___ + ___ + ___ + ___ + ___ = ___

___ × ___ = ___

Draw jumps on the number line to show equal groups. Find the product. (3.OA.A.3)

9. 6 groups of 3

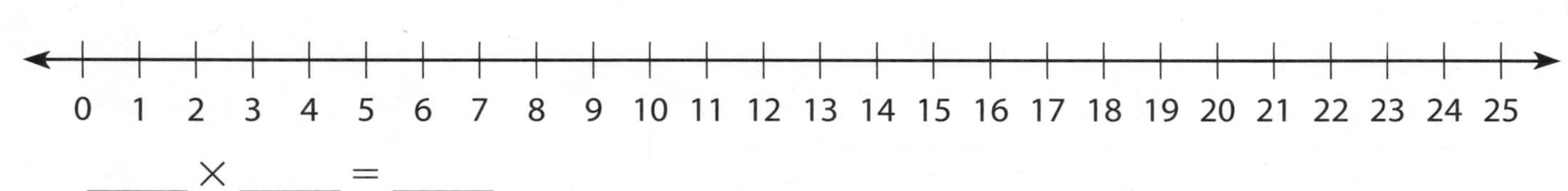

_____ × _____ = _____

10. Beth's mother cut some melons into equal slices. She put 4 slices each on 8 plates. Write a multiplication sentence to show the total number of melon slices she put on the plates. (3.OA.A.1)

____

11. Avery had 125 animal stickers. She gave 5 animal stickers to each of her 10 friends. How many animal stickers did she have left? What number sentences did you use to solve? (3.OA.A.3)

____

12. Matt made 2 equal groups of marbles. Write a multiplication sentence to show the total number of marbles. (3.OA.A.1)

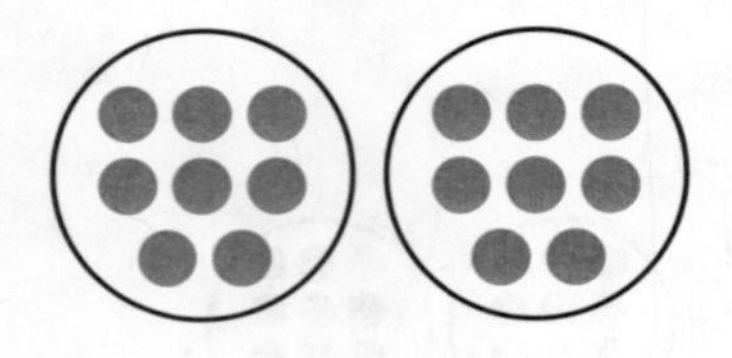

____

13. GO DEEPER Lindsey has 10 inches of ribbon. She buys another 3 lengths of ribbon, each 5 inches long. How much ribbon does she have now? (3.OA.A.3)

____

14. Jack's birthday is in 4 weeks. How many days is it until Jack's birthday? Describe how you could use a number line to solve. (3.OA.A.3)

____

____

____

Name ______________________________

Problem Solving • Model Multiplication

Essential Question How can you use the strategy *draw a diagram* to solve one- and two-step problems?

Common Core **Operations and Algebraic Thinking—3.OA.D.8** *Also 3.OA.A.1, 3.OA.A.3*

MATHEMATICAL PRACTICES
MP1, MP3, MP4, MP8

Unlock the Problem (Real World)

Three groups of students are taking drum lessons. There are 8 students in each group. How many students are taking drum lessons?

Read the Problem

What do I need to find?

I need to find how many ______________ are taking drum lessons.

What information do I need to use?

There are __________ groups of students taking drum lessons. There are __________ students in each group.

How will I use the information?

I will draw a bar model to help me see

____

_____________________________________.

Solve the Problem

Complete the bar model to show the drummers.

Write 8 in each box to show the 8 students in each of the 3 groups.

8	____	____

■ students

Since there are equal groups, I can multiply to find the number of students taking drum lessons.

____ × ____ = ■

____ = ■

So, there are ____ students in all.

Math Talk MATHEMATICAL PRACTICES 4

Use Models How would the bar model change if there were 6 groups of 4 students?

Try Another Problem

Twelve students in Mrs. Taylor's class want to start a band. Seven students each made a drum. The rest of the students made 2 shakers each. How many shakers were made?

Read the Problem	Solve the Problem
What do I need to find?	**Record the steps you used to solve the problem.**
What information do I need to use?	
How will I use the information?	

7	____

12 students

1. How many shakers in all did the students make? ____________________

2. How do you know your answer is reasonable? ____________________

Math Talk MATHEMATICAL PRACTICES 1

Evaluate Why wouldn't you draw 2 boxes and write 5 in each box?

Name ______________________

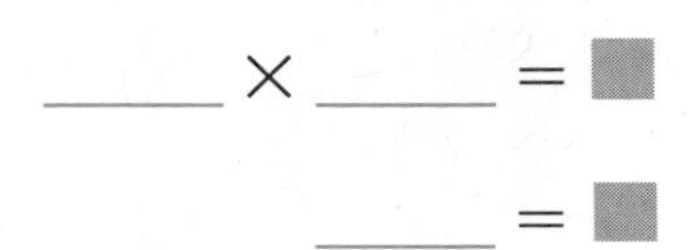

Share and Show

1. There are 6 groups of 4 students who play the trumpet in the marching band. How many students play the trumpet in the band?

First, draw a bar model to show each group of students.

Draw _____ boxes and write _____ in each box.

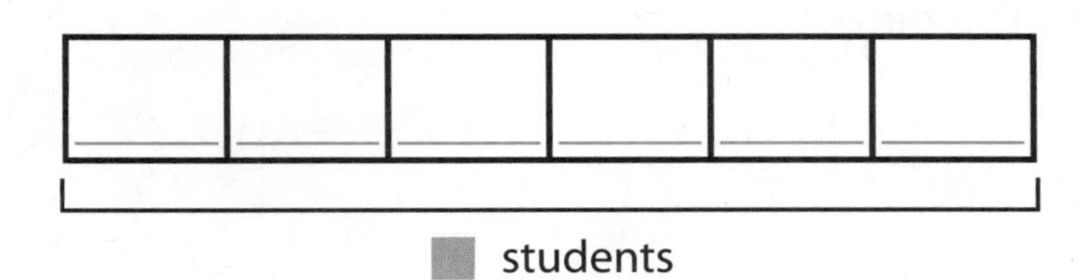

Then, multiply to find the total number of trumpet players.

_____ × _____ = ■

_____ = ■

So, _____ students play the trumpet in the marching band.

2. What if there are 4 groups of 7 students who play the saxophone? How many students play the saxophone or trumpet?

____

On Your Own

3. THINK SMARTER Suppose there are 5 groups of 4 trumpet players. In front of the trumpet players are 18 saxophone players. How many students play the trumpet or saxophone?

____

4. GO DEEPER In a garden there are 3 rows of plants. There are 5 plants in each row. Six of the plants are pumpkin plants and the rest are corn. How many corn plants are in the garden?

____

Use the picture graph for 5–7.

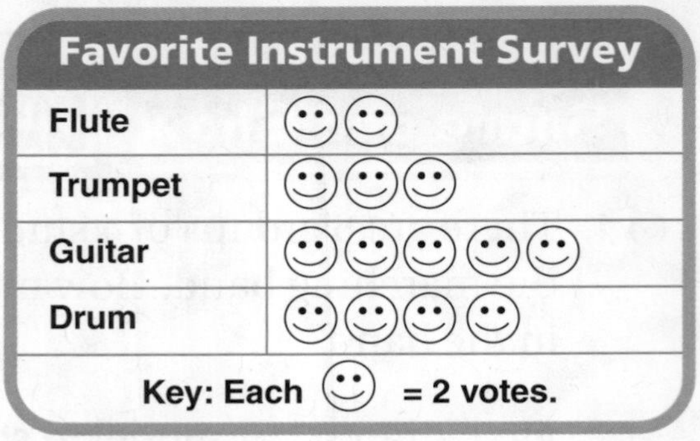

Favorite Instrument Survey	
Flute	☺☺
Trumpet	☺☺☺
Guitar	☺☺☺☺☺
Drum	☺☺☺☺
Key: Each ☺ = 2 votes.	

5. The picture graph shows how students in Jillian's class voted for their favorite instrument. How many students voted for the guitar?

____

6. GO DEEPER On the day of the survey, two students were absent. The picture graph shows the votes of all the other students in the class, including Jillian. How many students are in the class? Explain your answer.

____

____

____

7. THINK SMARTER Jillian added the number of votes for two instruments and got a total of 12 votes. For which two instruments did she add the votes?

_______________ and _______________

8. MATHEMATICAL PRACTICE 8 **Use Repeated Reasoning** The flute was invented 26 years after the harmonica. The electric guitar was invented 84 years after the flute. How many years was the electric guitar invented after the harmonica?

____

Personal Math Trainer

9. THINK SMARTER + Raul buys 4 packages of apple juice and 3 packages of grape juice. There are 6 drink boxes in each package. How many drink boxes does Raul buy? Show your work.

____

Name ______________________

Problem Solving • Model Multiplication

COMMON CORE STANDARD—3.OA.D.8
Solve problems involving the four operations, and identify and explain patterns in arithmetic.

Draw a diagram to solve each problem.

1. Robert put some toy blocks into 3 rows. There are 5 blocks in each row. How many blocks are there?

   15 blocks

2. Mr. Fernandez is putting tiles on his kitchen floor. There are 2 rows with 9 tiles in each row. How many tiles are there?

   ____

3. In Jillian's garden, there are 3 rows of carrots, 2 rows of string beans, and 1 row of peas. There are 8 plants in each row. How many plants are there in the garden?

   ____

4. Maya visits the movie rental store. On one wall, there are 6 DVDs on each of 5 shelves. On another wall, there are 4 DVDs on each of 4 shelves. How many DVDs are there on the shelves?

   ____

5. The media center at Josh's school has a computer area. The first 4 rows have 6 computers each. The fifth row has 4 computers. How many computers are there?

   ____

6. **WRITE** *Math* Describe one kind of diagram you might draw to help you solve a problem.

   ____

   ____

Lesson Check (3.OA.D.8)

1. There are 5 shelves of video games in a video store. There are 6 video games on each shelf. How many video games are there on the shelves?

2. Ken watches a marching band. He sees 2 rows of flute players. Six people are in each row. He sees 8 trombone players. How many flute players and trombone players does Ken see?

Spiral Review (3.NBT.1, 3.NBT.2, 3.MD.3)

3. What is the sum of 438 and 382?

4. Estimate the sum.

$$\begin{array}{r} 622 \\ +\ 84 \\ \hline \end{array}$$

5. Francine uses 167 silver balloons and 182 gold balloons for her store party. How many silver and gold balloons does Francine use?

6. Yoshi is making a picture graph. Each picture of a soccer ball stands for two goals he scored for his team. The row for January has 9 soccer balls. How many goals did Yoshi score during January?

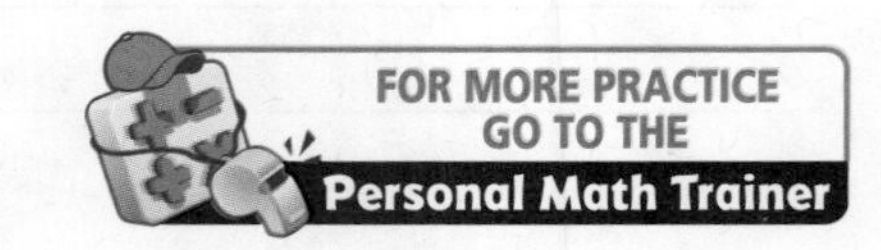

Name ______________________

Model with Arrays

Essential Question How can you use arrays to model multiplication and find factors?

Common Core **Operations and Algebraic Thinking—3.OA.A.3** *Also 3.OA.A.1*

MATHEMATICAL PRACTICES
MP2, MP4, MP5

Unlock the Problem

Real World

Hands On

Many people grow tomatoes in their gardens. Lee plants 3 rows of tomato plants with 6 plants in each row. How many tomato plants are there?

▲ Tomatoes are a great source of vitamins.

Activity 1

Materials ■ square tiles ■ MathBoard

- You make an **array** by placing the same number of tiles in each row. Make an array with 3 rows of 6 tiles to show the tomato plants.
- Now draw the array you made.

- Find the total number of tiles.

Multiply. $3 \times 6 =$ _____

3 ↑ number of rows; 6 ↑ number in each row

So, there are _____ tomato plants.

Math Talk MATHEMATICAL PRACTICES 2

Reason Abstractly Does the number of tiles change if you turn the array to show 6 rows of 3?

Activity 2 **Materials** ■ square tiles ■ MathBoard

Use 8 tiles. Make as many different arrays as you can, using all 8 tiles. Draw the arrays. The first one is done for you.

A

1 row of 8

$1 \times 8 = 8$

B

8 rows of ____

$8 \times$ ____ $= 8$

C

____ rows of ____

____ $\times$ ____ $= 8$

D

____ rows of ____

____ $\times$ ____ $= 8$

You can make ____ different arrays using 8 tiles.

Share and Show

1. Complete. Use the array.

____ rows of ____ = ____

____ $\times$ ____ = ____

Write a multiplication sentence for the array.

✓2.

____

✓3.

____

Name ______________________

On Your Own

Write a multiplication sentence for the array.

4. 

____

5. 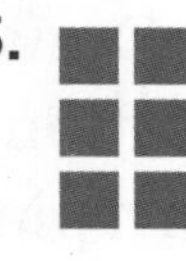

____

Draw an array to find the product.

6. $3 \times 6 =$ _____

7. $4 \times 7 =$ _____

8. GO DEEPER DeShawn makes an array using 3 rows of 5 tiles. How many tiles does Deshawn have if he adds 2 more rows to the array?

____

9. GO DEEPER Ming makes an array using 2 rows of 7 tiles. She adds 3 more rows to the array. Write a multiplication sentence that shows Ming's array.

____

10. GO DEEPER Use 6 tiles. Make as many different arrays as you can using all the tiles. Draw the arrays. Then write a multiplication sentence for each array.

____

____

Problem Solving • Applications Real World

Use the table to solve 11–12.

Mr. Bloom's Garden

Vegetable	Planted In
Beans	4 rows of 6
Carrots	2 rows of 8
Corn	5 rows of 9
Beets	4 rows of 7

11. MATHEMATICAL PRACTICE 4 **Use Models** Mr. Bloom grows vegetables in his garden. Draw an array and write the multiplication sentence to show how many corn plants Mr. Bloom has in his garden.

____

12. THINKSMARTER Could Mr. Bloom have planted his carrots in equal rows of 4? If so, how many rows could he have planted? Explain.

____

____

____

____

13. MATHEMATICAL PRACTICE 5 **Communicate** Mr. Bloom has 12 strawberry plants. Describe all of the different arrays that Mr. Bloom could make using all of his strawberry plants. The first one is done for you.

2 rows of 6; _______________

____

14. THINKSMARTER Elizabeth ran 3 miles each day for 5 days. How many miles did she run in all? Shade the array to represent the problem. Then solve.

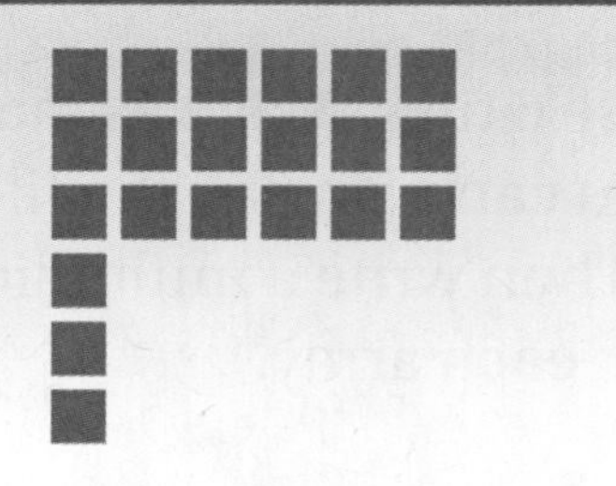

____

Name ______________________

Model with Arrays

COMMON CORE STANDARD—3.OA.A.3
Represent and solve problems involving multiplication and division.

Write a multiplication sentence for the array.

1. 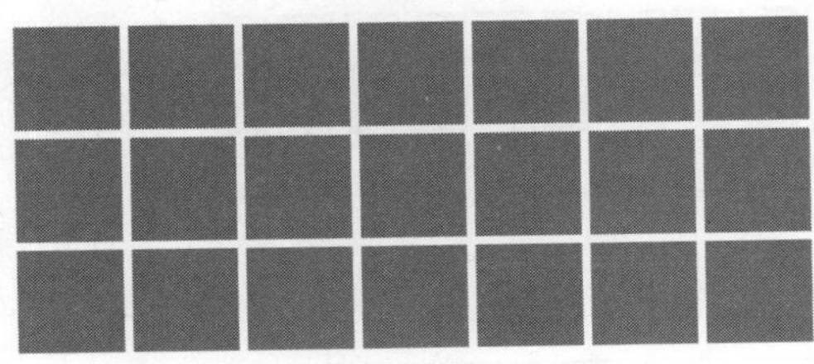

$3 \times 7 =$ 21

2. 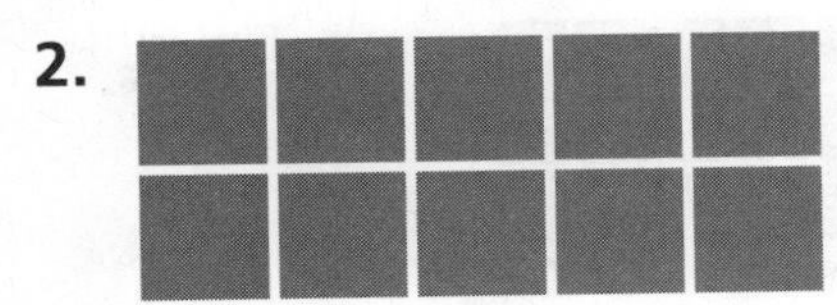

$2 \times 5 =$ _____

Draw an array to find the product.

3. $4 \times 2 =$ _____

4. $2 \times 8 =$ _____

Problem Solving Real World

5. Lenny is moving tables in the school cafeteria. He places all the tables in a 7×4 array. How many tables are in the cafeteria?

____

6. Ms. DiMeo directs the school choir. She has the singers stand in 3 rows. There are 8 singers in each row. How many singers are there?

____

7. WRITE ▸ *Math* Write a word problem that can be solved by drawing an array. Then draw the array and solve the problem.

____

____

Lesson Check (3.OA.A.3)

1. What multiplication sentence does this array show?

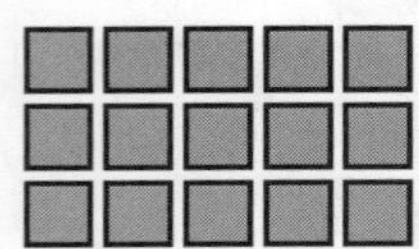

____

2. What multiplication sentence does this array show?

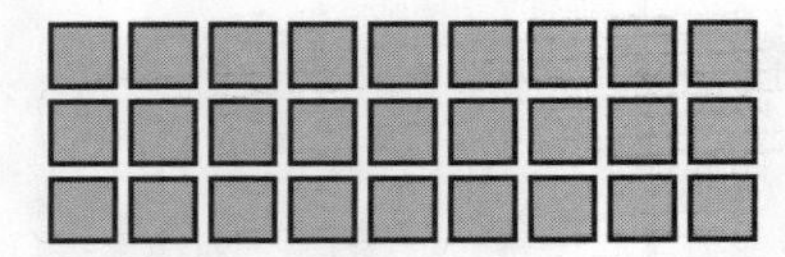

____

Spiral Review (3.NBT.A.1, 3.NBT.A.2, 3.MD.B.3)

3. Use the table to find who traveled 700 miles farther than Paul during summer vacation.

Summer Vacations	
Name	**Distance in Miles**
Paul	233
Andrew	380
Bonnie	790
Tara	933
Susan	853

____

4. Use the bar graph to find what hair color most students have.

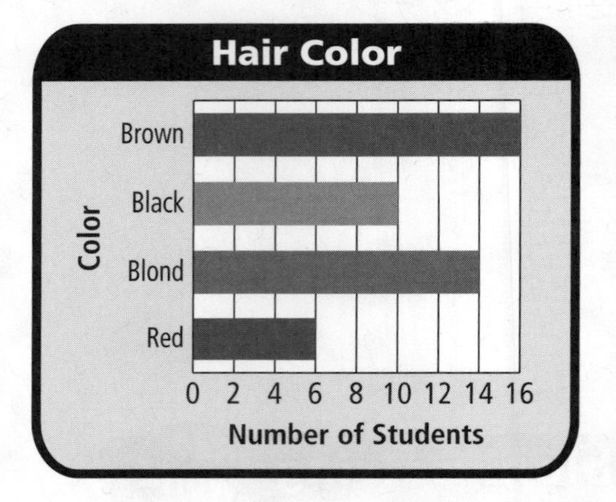

____

5. Spencer orders 235 cans of tomatoes to make salsa for the festival. What is 235 rounded to the nearest ten?

____

6. Which bar would be the longest on a bar graph of the data?

Favorite Pizza Topping	
Topping	**Votes**
Cheese	5
Pepperoni	4
Vegetable	1
Sausage	3

____

FOR MORE PRACTICE GO TO THE **Personal Math Trainer**

Name ___________________________

Commutative Property of Multiplication

Essential Question How can you use the Commutative Property of Multiplication to find products?

Operations and Algebraic Thinking—3.OA.B.5 *Also 3.OA.A.1, 3.OA.A.3, 3.OA.C.7*

MATHEMATICAL PRACTICES
MP1, MP2, MP6, MP7

Unlock the Problem

Real World

Hands On

Dave works at the Bird Store. He arranges 15 boxes of birdseed in rows on the shelf. What are two ways he can arrange the boxes in equal rows?

- Circle the number that is the product.

Activity Make an array.

Materials ■ square tiles ■ MathBoard

Arrange 15 tiles in 5 equal rows.
Draw a quick picture of your array.

How many tiles are in each row? __________

What multiplication sentence does your array show? ______________

Suppose Dave arranges the boxes in 3 equal rows.
Draw a quick picture of your array.

How many tiles are in each row? __________

What multiplication sentence does your array show?

____

So, two ways Dave can arrange the 15 boxes are

in _____ rows of 3 or in 3 rows of _____.

MATHEMATICAL PRACTICES 7

Identify Relationships When using an array to help solve a multiplication problem, why does the answer stay the same when the array is turned?

Multiplication Property The **Commutative Property of Multiplication** states that when you change the order of the factors, the product stays the same. You can think of it as the Order Property of Multiplication.

$2 \times$ _____ $=$ _____

$3 \times$ _____ $=$ _____

Math Idea

Facts that show the Commutative Property of Multiplication have the same factors in a different order.

$2 \times 3 = 6$ and $3 \times 2 = 6$

So, $2 \times$ _____ $= 3 \times$ _____.

- Explain how the models are alike and how they are different.

____

____

Try This! **Draw a quick picture on the right that shows the Commutative Property of Multiplication. Then complete the multiplication sentences.**

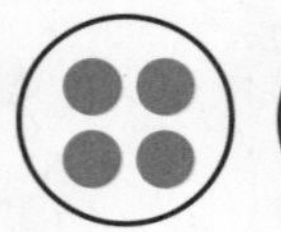
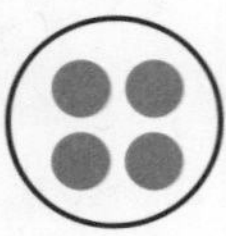

_____ $\times 4 =$ _____   _____ $\times 3 =$ _____

B

$2 \times$ _____ $=$ _____ $5 \times$ _____ $=$ _____

Name ________________________________

Share and Show

1. Write a multiplication sentence for the array.

____

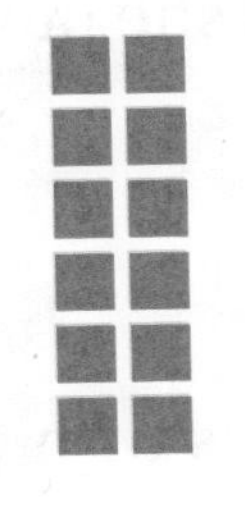

____

Math Talk MATHEMATICAL PRACTICES 1

Make Sense of Problems Explain what the factor 2 means in each multiplication sentence.

Write a multiplication sentence for the model. Then use the Commutative Property of Multiplication to write a related multiplication sentence.

2.

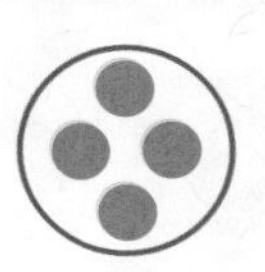
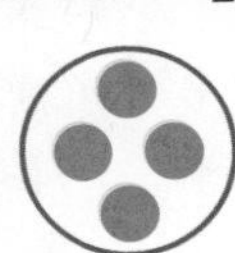

___ × ___ = ___

___ × ___ = ___

✓3.

___ × ___ = ___

___ × ___ = ___

✓4.

___ × ___ = ___

___ × ___ = ___

On Your Own

Write a multiplication sentence for the model. Then use the Commutative Property of Multiplication to write a related multiplication sentence.

5.

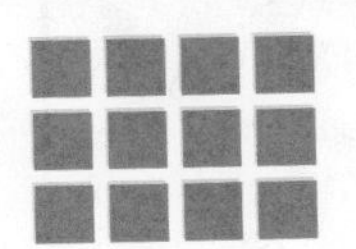

___ × ___ = ___

___ × ___ = ___

6.

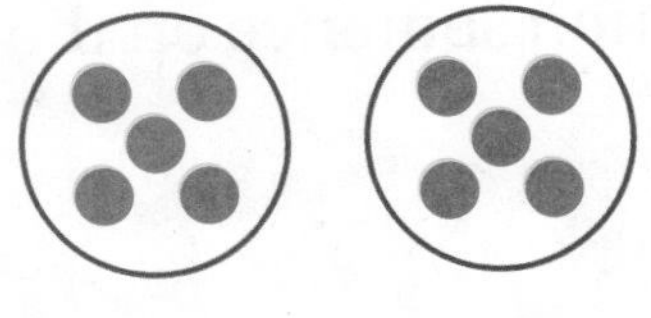

___ × ___ = ___

___ × ___ = ___

7.

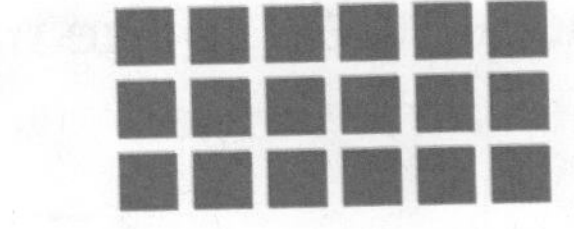

___ × ___ = ___

___ × ___ = ___

Use Reasoning **Algebra** **Write the unknown factor.**

8. 3 × 7 = ______ × 3

9. 4 × 5 = 10 × ______

10. 3 × 6 = ______ × 9

11. 6 × ______ = 4 × 9

12. ______ × 8 = 4 × 6

13. 5 × 8 = 8 × ______

Problem Solving • Applications

14. Jenna used pinecones to make 18 peanut butter bird feeders. She hung the same number of feeders in 6 trees. Draw an array to show how many feeders she put in each tree.

She put _____ bird feeders in each tree.

15. GO DEEPER Mr. Diaz sets out 6 rows of glasses with 3 glasses in each row. Mrs. Diaz sets out 3 rows of glasses with 6 glasses in each row. How many glasses do Mr. and Mrs. Diaz set out in all?

____

16. GO DEEPER Write two different word problems about 12 birds to show 2×6 and 6×2. Solve each problem.

____

____

____

Math on the Spot

17. THINK SMARTER There are 4 rows of 6 bird stickers in Don's sticker album. There are 7 rows of 5 bird stickers in Lindsey's album. How many bird stickers do they have?

____

18. THINK SMARTER Write the letter for each multiplication sentence on the left next to the multiplication sentence on the right that has the same value.

Ⓐ $5 \times 7 = ■$ ☐ $6 \times 3 = ■$

Ⓑ $8 \times 2 = ■$ ☐ $2 \times 8 = ■$

Ⓒ $3 \times 6 = ■$ ☐ $4 \times 9 = ■$

Ⓓ $9 \times 4 = ■$ ☐ $7 \times 5 = ■$

Name ____________________

Commutative Property of Multiplication

Common Core

COMMON CORE STANDARD—3.OA.B.5
Understand properties of multiplication and the relationship between multiplication and division.

Write a multiplication sentence for the model. Then use the Commutative Property of Multiplication to write a related multiplication sentence.

1\.

5 × 2 = 10

2 × 5 = 10

2\.

____ × ____ = ____

____ × ____ = ____

3\.

____ × ____ = ____

____ × ____ = ____

4\.

____ × ____ = ____

____ × ____ = ____

Problem Solving Real World

5\. A garden store sells trays of plants. Each tray holds 2 rows of 8 plants. How many plants are in one tray?

____

6\. Jeff collects toy cars. They are displayed in a case that has 4 rows. There are 6 cars in each row. How many cars does Jeff have?

____

7\. WRITE Math How are the Commutative Property of Addition and the Commutative Property of Multiplication alike?

____

____

Lesson Check (3.OA.B.5)

1. Write a sentence that shows the Commutative Property of Multiplication.

2. What factor makes the number sentence true?

   $7 \times 4 = \blacksquare \times 7$

Spiral Review (3.NBT.A.1, 3.NBT.A.2, 3.MD.B.3)

3. Ms. Williams drove 149 miles on Thursday and 159 miles on Friday. About how many miles did she drive altogether?

4. Inez has 699 pennies and 198 nickels. Estimate how many more pennies than nickels she has.

5. This year, the parade had 127 floats. That was 34 fewer floats than last year. How many floats were in the parade last year?

6. Jeremy made a tally table to record how his friends voted for their favorite pet. His table shows 卌 卌 || next to Dog. How many friends voted for dog?

FOR MORE PRACTICE GO TO THE **Personal Math Trainer**

Name ______________________

ALGEBRA
Lesson 3.7

Multiply with 1 and 0

Essential Question What happens when you multiply a number by 0 or 1?

Operations and Algebraic Thinking—3.OA.B.5 *Also 3.OA.A.1, 3.OA.A.3, 3.OA.C.7*

MATHEMATICAL PRACTICES
MP1, MP2, MP3, MP6

Unlock the Problem

Luke sees 4 birdbaths. Each birdbath has 2 birds in it. What multiplication sentence tells how many birds there are?

- How many birdbaths are there? ______________
- How many birds does Luke see in each birdbath? ______________

Draw a quick picture to show the birds in the birdbaths.

_____ × _____ = _____

One bird flies away from each birdbath. Cross out 1 bird in each birdbath above. What multiplication sentence shows the total number of birds now?

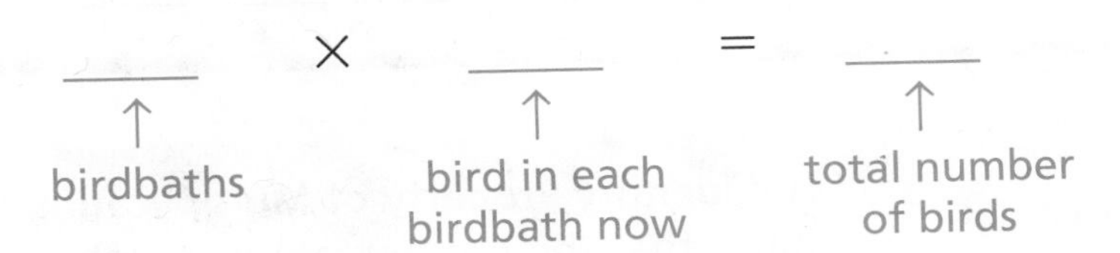

Now cross out another bird in each birdbath. What multiplication sentence shows the total number of birds in the birdbaths now?

- How do the birdbaths look now? ______________

Math Talk MATHEMATICAL PRACTICES 1

Analyze What if there were 5 birdbaths with 0 birds in each of them? What would be the product? Explain.

Example

Jenny has 2 pages of bird stickers. There are 4 stickers on each page. How many stickers does she have in all?

$2 \times 4 =$ ____ **Think:** 2 groups of 4

So, Jenny has ____ stickers in all.

Suppose Jenny uses 1 page of the stickers. What fact shows how many stickers she has now?

____ $\times$ ____ $=$ ____ **Think:** 1 group of 4

So, Jenny has ____ stickers now.

Then, Jenny uses the rest of the stickers. What fact shows how many stickers Jenny has now?

____ $\times$ ____ $=$ ____ **Think:** 0 groups of 4

So, Jenny has ____ stickers now.

ERROR Alert

A 0 in a multiplication sentence means 0 groups or 0 things in a group, so the product is always 0.

- What does each number in $0 \times 4 = 0$ tell you?

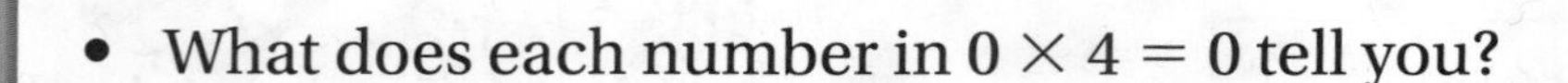

____

1. What pattern do you see when you multiply numbers with 1 as a factor?

Think: $1 \times 2 = 2$ $1 \times 3 = 3$ $1 \times 4 = 4$

____

____

The **Identity Property of Multiplication** states that the product of any number and 1 is that number.

$7 \times 1 = 7$ $6 \times 1 = 6$

$1 \times 7 = 7$ $1 \times 6 = 6$

2. What pattern do you see when you multiply numbers with 0 as a factor?

Think: $0 \times 1 = 0$ $0 \times 2 = 0$ $0 \times 5 = 0$

____

____

The **Zero Property of Multiplication** states that the product of zero and any number is zero.

$0 \times 5 = 0$ $0 \times 8 = 0$

$5 \times 0 = 0$ $8 \times 0 = 0$

Name ______________________________

Share and Show

1. What multiplication sentence matches this picture? Find the product.

____

Find the product.

2. $5 \times 1 =$ ____
3. $0 \times 2 =$ ____
4. $4 \times 0 =$ ____
5. $1 \times 6 =$ ____
6. $3 \times 0 =$ ____
7. $1 \times 2 =$ ____
8. $0 \times 6 =$ ____
9. $8 \times 1 =$ ____

Math Talk

MATHEMATICAL PRACTICES 6

Compare Explain how 3×1 and $3 + 1$ are different.

On Your Own

Find the product.

10. $3 \times 1 =$ ____
11. $8 \times 0 =$ ____
12. $1 \times 9 =$ ____
13. $0 \times 7 =$ ____

MATHEMATICAL PRACTICE 2 **Use Reasoning** **Algebra** **Complete the multiplication sentence.**

14. ____ $\times 1 = 15$
15. $1 \times 28 =$ ____
16. $0 \times 46 =$ ____
17. $36 \times 0 =$ ____
18. ____ $\times 5 = 5$
19. $19 \times$ ____ $= 0$
20. ____ $\times 0 = 0$
21. $7 \times$ ____ $= 7$

22. Noah sets out 7 baskets at the Farmers' Market. Each basket holds 1 watermelon. How many watermelons does Noah set out?

____

23. GO DEEPER Mason and Alexis each have 1 bag of marbles. There are 9 marbles in each bag. How many marbles do they have altogether?

____

24. GO DEEPER Each box holds 6 black markers and 4 red markers. Derek has 0 boxes of markers. Write a number sentence that shows how many markers Derek has. Explain how you found your answer.

____

____

Problem Solving • Applications

Use the table for 25–27.

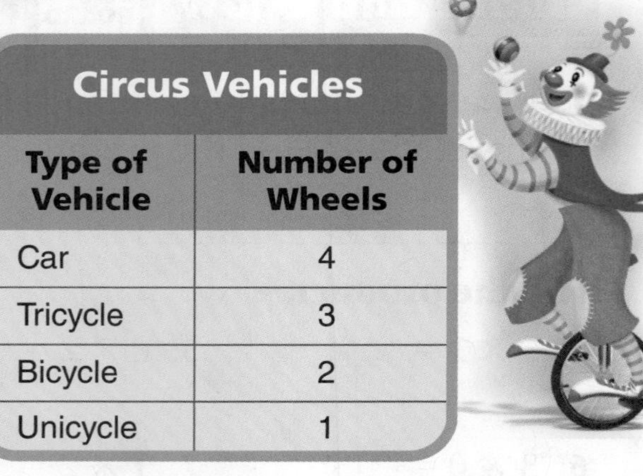

Circus Vehicles	
Type of Vehicle	**Number of Wheels**
Car	4
Tricycle	3
Bicycle	2
Unicycle	1

25. At the circus Jon saw 5 unicycles. How many wheels are on the 5 unicycles? Write a multiplication sentence.

_____ × _____ = _________

26. **What's the Question?** Julia used multiplication with 1 and the information in the table. The answer is 3.

____

____

27. THINK SMARTER Brian saw some circus vehicles. He saw 17 wheels in all. If 2 of the vehicles are cars, how many vehicles are bicycles and tricycles?

____

28. WRITE ▸ Math Write a word problem that uses multiplying with 1 or 0. Show how to solve your problem.

____

____

29. THINK SMARTER For numbers 29a–29d, select True or False for each multiplication sentence.

29a. $6 \times 0 = 0$ ○ True ○ False

29b. $0 \times 9 = 9 \times 0$ ○ True ○ False

29c. $1 \times 0 = 1$ ○ True ○ False

29d. $3 \times 1 = 3$ ○ True ○ False

Name ______

Multiply with 1 and 0

Common Core

COMMON CORE STANDARD—3.OA.B.5
Understand properties of multiplication and the relationship between multiplication and division.

Find the product.

1. $1 \times 4 =$ 4
2. $0 \times 8 =$ ____
3. $0 \times 4 =$ ____
4. $1 \times 6 =$ ____
5. $3 \times 0 =$ ____
6. $0 \times 9 =$ ____
7. $8 \times 1 =$ ____
8. $1 \times 2 =$ ____
9. $10 \times 1 =$ ____
10. $2 \times 0 =$ ____
11. $5 \times 1 =$ ____
12. $1 \times 0 =$ ____
13. $0 \times 0 =$ ____
14. $1 \times 3 =$ ____
15. $9 \times 0 =$ ____
16. $1 \times 1 =$ ____

Problem Solving

17. Peter is in the school play. His teacher gave 1 copy of the play to each of 6 students. How many copies of the play did the teacher hand out?

______

18. There are 4 egg cartons on the table. There are 0 eggs in each carton. How many eggs are there in all?

______

19. **WRITE** *Math* One group has 5 people, and each person has 1 granola bar. Another group has 5 people, and each person has 0 granola bars. Which group has more granola bars? Explain.

______

______

______

Lesson Check (3.OA.B.5)

1. There are 0 bicycles in each bicycle rack. If there are 8 bicycle racks, how many bicycles are there in the rack?

2. What is the product?

$1 \times 0 =$ ___

Spiral Review (3.NBT.A.2, 3.OA.A.3, 3.MD.B.3)

3. Mr. Ellis drove 197 miles on Monday and 168 miles on Tuesday. How many miles did he drive?

4. What multiplication sentence does the array show?

Use the bar graph for 5–6.

5. How many cars were washed on Friday and Saturday combined?

6. How many more cars were washed on Saturday than on Sunday?

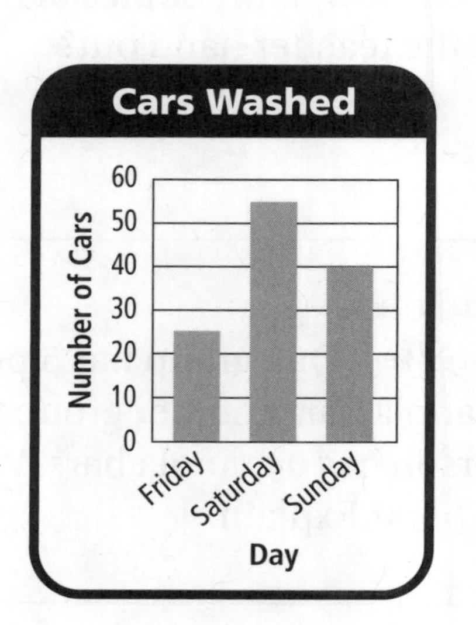

FOR MORE PRACTICE GO TO THE **Personal Math Trainer**

Name ______________________________

Chapter 3 Review/Test

1. There are 3 boats on the lake. Six people ride in each boat. How many people ride in the boats? Draw circles to model the problem and explain how to solve it.

_______ people

2. Nadia has 4 sheets of stickers. There are 8 stickers on each sheet. She wrote this number sentence to represent the total number of stickers.

$$4 \times 8 = 32$$

What is a related number sentence that also represents the total number of stickers she has?

(A) $8 + 4 = \blacksquare$

(B) $4 + 4 + 4 + 4 = \blacksquare$

(C) $8 \times 8 = \blacksquare$

(D) $8 \times 4 = \blacksquare$

3. Lindsay went hiking for two days in Yellowstone National Park. The first jump on the number line shows how many birds she saw the first day. She saw the same number of birds the next day.

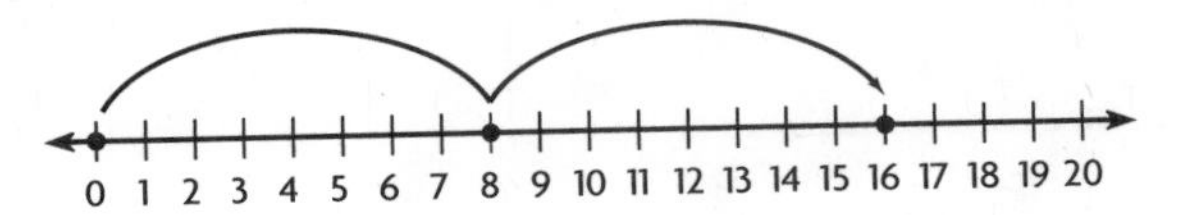

Write the multiplication sentence that is shown on the number line.

_______ × _______ = _______

GO DIGITAL Assessment Options
Chapter Test

4. Paco drew an array to show the number of desks in his classroom.

Write a multiplication sentence for the array.

____

5. Alondra makes 4 necklaces. She uses 5 beads on each necklace.

For numbers 5a–5d, choose Yes or No to tell if the number sentence could be used to find the number of beads Alondra uses.

5a. $4 \times 5 = \blacksquare$ ○ Yes ○ No

5b. $4 + 4 + 4 + 4 = \blacksquare$ ○ Yes ○ No

5c. $5 + 5 + 5 + 5 = \blacksquare$ ○ Yes ○ No

5d. $5 + 4 = \blacksquare$ ○ Yes ○ No

6. John sold 3 baskets of apples at the market. Each basket contained 9 apples. How many apples did John sell? Make a bar model to solve the problem.

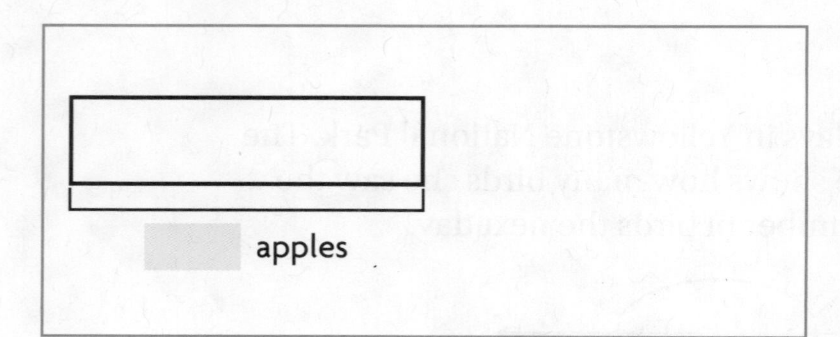

Name ______________________________

7. Select the number sentences that show the Commutative Property of Multiplication. Mark all that apply.

Ⓐ $3 \times 2 = 2 \times 3$

Ⓑ $4 \times 9 = 4 \times 9$

Ⓒ $5 \times 0 = 0$

Ⓓ $6 \times 1 = 1 \times 6$

Ⓔ $7 \times 2 = 14 \times 1$

8. A waiter carried 6 baskets with 5 dinner rolls in each basket. How many dinner rolls did he carry? Show your work.

__________ dinner rolls

9. Sonya needs 3 equal lengths of wire to make 3 bracelets. The jump on the number line shows the length of one wire in inches. How many inches of wire will Sonya need to make the 3 bracelets?

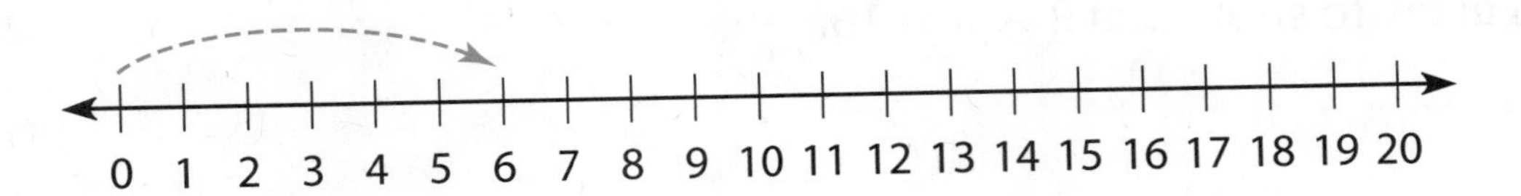

__________ inches

10. Josh has 4 dogs. Each dog gets 2 dog biscuits every day. How many biscuits will Josh need for all of his dogs for Saturday and Sunday?

__________ biscuits

11. GO DEEPER Jorge displayed 28 cans of paint on a shelf in his store.

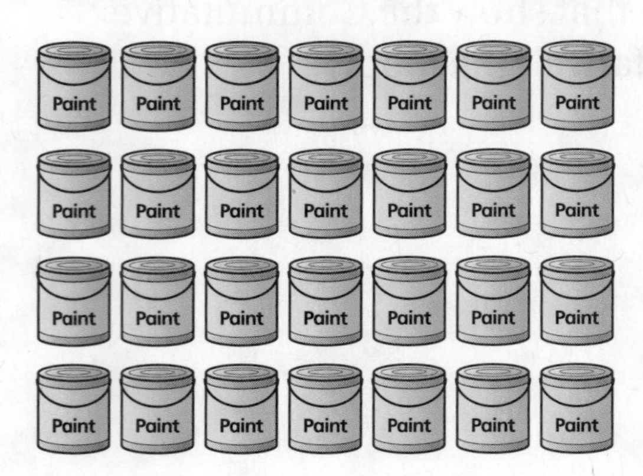

Select other ways Jorge could arrange the same number of cans. Mark all that apply.

- (A) 2 rows of 14
- (B) 1 row of 28
- (C) 6 rows of 5
- (D) 8 rows of 3
- (E) 7 rows of 4

12. Choose the number that makes the statement true.

The product of any number and [0 / 1 / 10] is zero.

13. James made this array to show that $3 \times 5 = 15$.

Part A

James says that $5 \times 3 = 15$. Is James correct? Draw an array to explain your answer.

____

Part B

Which number property supports your answer?

____

____

Name ______________________________

14. Julio has a collection of coins. He puts the coins in 2 equal groups. There are 6 coins in each group. How many coins does Julio have? Use the number line to show your work.

__________ coins

15. GO DEEPER Landon collects trading cards.

Part A

Yesterday, Landon sorted his trading cards into 4 groups. Each group had 7 cards. Draw a bar model to show Landon's cards. How many cards does he have?

__________ trading cards

Part B

Landon buys 3 more packs of trading cards today. Each pack has 8 cards. Write a multiplication sentence to show how many cards Landon buys today. Then find how many cards Landon has now. Show your work.

____

16. A unicycle has only 1 wheel. Write a multiplication sentence to show how many wheels there are on 9 unicycles.

__________ $\times$ __________ $=$ __________

17. Carlos spent 5 minutes working on each of 8 math problems. He can use 8×5 to find the total amount of time he spent on the problems.

For numbers 17a–17d, choose Yes or No to show which are equal to 8×5.

17a. $8 + 5$ ○ Yes ○ No

17b. $5 + 5 + 5 + 5 + 5$ ○ Yes ○ No

17c. $8 + 8 + 8 + 8 + 8$ ○ Yes ○ No

17d. $5 + 5 + 5 + 5 + 5 + 5 + 5 + 5$ ○ Yes ○ No

18. Lucy and her mother made tacos. They put 2 tacos on each of 7 plates.

Select the number sentences that show all the tacos Lucy and her mother made. Mark all that apply.

Ⓐ $2 + 2 + 2 + 2 + 2 + 2 + 2 = 14$

Ⓑ $2 + 7 = 9$

Ⓒ $7 + 7 = 14$

Ⓓ $8 + 6 = 14$

Ⓔ $2 \times 7 = 14$

Personal Math Trainer

19. THINK SMARTER + Jayson is making 5 sock puppets. He glues 2 buttons on each puppet for its eyes. He glues 1 pompom on each puppet for its nose.

Part A

Write the total number of buttons and pompoms he uses. Write a multiplication sentence for each.

Eyes

_____ buttons

_____ × _____ = _____

Noses

_____ pompoms

_____ × _____ = _____

Part B

After making 5 puppets, Jayson has 4 buttons and 3 pompoms left. What is the greatest number of puppets he can make with those items if he wants all his puppets to look the same? Draw models and use them to explain.

At most, he can make _____ more puppets.

____

____

____

____